Cambridge Elements ≡

Elements in the Philosophy of Mathematics
edited by
Penelope Rush
University of Tasmania
Stewart Shapiro
The Ohio State University

A CONCISE HISTORY OF MATHEMATICS FOR PHILOSOPHERS

John Stillwell
University of San Francisco

CAMBRIDGE
UNIVERSITY PRESS

CAMBRIDGE
UNIVERSITY PRESS

University Printing House, Cambridge CB2 8BS, United Kingdom

One Liberty Plaza, 20th Floor, New York, NY 10006, USA

477 Williamstown Road, Port Melbourne, VIC 3207, Australia

314–321, 3rd Floor, Plot 3, Splendor Forum, Jasola District Centre,
New Delhi – 110025, India

79 Anson Road, #06–04/06, Singapore 079906

Cambridge University Press is part of the University of Cambridge.

It furthers the University's mission by disseminating knowledge in the pursuit of
education, learning, and research at the highest international levels of excellence.

www.cambridge.org
Information on this title: www.cambridge.org/9781108456234
DOI: 10.1017/9781108610124

First published 2019

A catalogue record for this publication is available from the British Library.

ISBN 978-1-108-45623-4 Paperback
ISSN 2399-2883 (online)
ISSN 2514-3808 (print)

A Concise History of Mathematics for Philosophers

Elements in the Philosophy of Mathematics

DOI: 10.1017/9781108610124
First published online: June 2019

John Stillwell
University of San Francisco
Author for correspondence: John Stillwell, stillwell@usfca.edu

Abstract: This Element aims to present an outline of mathematics and its history, with particular emphasis on events that shook up its philosophy. It ranges from the discovery of irrational numbers in ancient Greece to the nineteenth- and twentieth-century discoveries on the nature of infinity and proof. Recurring themes are intuition and logic, meaning and existence, and the discrete and the continuous. These themes have evolved under the influence of new mathematical discoveries, and the story of their evolution is, to a large extent, the story of philosophy of mathematics.

Keywords: mathematics, philosophy

ISBNs: 9781108456234 (PB), 9781108610124 (OC)
ISSNs: 2399-2883 (online), 2514-3808 (print)

Contents

Preface

Since ancient times, there has been a struggle between mathematics and its philosophy. As soon as there seems to be a settled view of the nature of mathematics, some new mathematical discovery comes along to disrupt it. Thus, the Pythagorean view that 'all is number' was disrupted by the discovery of irrational lengths, and the philosophy of mathematics had to expand to include a separate field of geometry. But this raised the question, Can the geometric view be reconciled with the numerical view? If so, how? And so it went, for millennia.

In many cases, advances *in* mathematics changed ideas *about* mathematics, by forcing the acceptance of concepts previously thought impossible or paradoxical. Thus mathematics disrupted philosophy. In the opposite direction, philosophy kept mathematics honest by pointing out contradictions and suggesting how concepts might be clarified in order to resolve them. Sometimes the philosopher and the mathematician were one and the same person – such as Descartes, Leibniz, or Bolzano – so one might almost say that mathematics is an especially rich and stable branch of philosophy. At any rate, if one is to understand the past and present state of the philosophy of mathematics, one must first understand mathematics, and its history.

The aim of the present Element is to give a brief introduction to mathematics and its history, with particular emphasis on events that shook up its philosophy. If you like, it is a book on 'mathematics for philosophers'. I try not to take a particular philosophical position, except to say that I believe that mathematics guides philosophy, more so than the other way round. As a corollary, I believe that mathematicians have made important contributions to philosophy, even when it was not their intention.

Each section begins with a preview of topics to be discussed and ends with a section highlighting the philosophical questions raised by the mathematics. The same themes recur from section to section – intuition and logic, meaning and existence, and the discrete and the continuous – but they evolve under the influence of new mathematical discoveries.

Experts may be surprised that there is little or no mention of philosophies of mathematics that were prominent in the twentieth century – platonism, logicism, formalism, nominalism, and intuitionism, for example. This is partly because I find none of them adequate, but mainly because I hope to look at the philosophy of mathematics without being influenced by labels. I want to present as much *philosophically instructive mathematics* as possible and leave readers to decide how it should be sorted and labelled in philosophical terms. My hope is that this Element will equip readers with a 'mathematical lens' with which to view many philosophical issues.

I thank Jeremy Avigad, Rossella Lupacchini, Wilfried Sieg, and an anonymous referee for their helpful comments, which have resulted in many improvements.

1 Irrational Numbers and Geometry

PREVIEW

The source of many issues in the philosophy of mathematics – the nature of proof and truth; the meaning and existence of numbers; the role of infinity; and the relation between geometry, algebra, and arithmetic – is Euclid's *Elements* from around 300 BCE. The *Elements* is best known for its axiomatic geometry – Euclidean geometry – which includes proofs of signature results such as the Pythagorean theorem and the existence of exactly five regular polyhedra. However, the *Elements* also includes fundamental ideas of number theory, such as the existence of infinitely many prime numbers, the Euclidean algorithm for greatest common divisor, and (an equivalent of) unique prime factorization.

In Euclid's time, as now, there was a conceptual gulf between geometry and number theory – between measuring and counting, or between the continuous and the discrete. The major reason for this gulf was the existence of irrationals, discovered before Euclid's time by the Pythagoreans and, by the time of the *Elements*, the subject of a sophisticated 'theory of proportions'. This theory, in Book V of the *Elements*, made a tenuous bridge between the continuous and the discrete. The bridge was gradually strengthened over the centuries by the work of later mathematicians, but not without philosophical conflicts and mathematical surprises.

These issues are the subject of this section and the next.

1.1 The Pythagorean Theorem

The Pythagorean theorem was discovered independently several times in human history, and in several different cultures. So if any theorem typifies mathematics – and its universality – this is it. Figure 1 illustrates the theorem: the (grey) square on the hypotenuse of the (white) right-angled triangle is equal to sum of the (black) squares on the other two sides.

Figure 2 shows a plausible 'proof by picture' of the theorem: the grey square equals the big square minus four copies of the triangle, which in turn equals the sum of the two black squares.

Most of the independent discoveries of the theorem were probably like this, and indeed the human visual system has many mathematical discoveries to its credit. Nevertheless, it was the radically different *axiomatic* path to theorems,

Figure 1 The Pythagorean theorem.

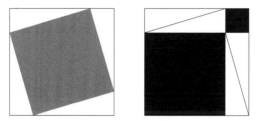

Figure 2 Seeing the Pythagorean theorem.

discussed in Section 1.4, that set the direction of mathematics for the next 2000 years.

But before the axiomatic path was established, the Pythagorean theorem provoked another important conceptual development: a distinction between length and number. Legend has it that the philosophy of the Pythagoreans was 'all is number', prompted by the discovery that whole number ratios govern musical harmony. This philosophy was overturned when irrational ratios were found in geometry – because of the Pythagorean theorem.

1.2 Irrationality

The Pythagorean theorem talks about sums of squares – an operation we will say more about below – but indirectly, it also tells us something about lengths. In particular, it says that if a triangle has perpendicular sides of length 1, then its hypotenuse has the length l whose square is 2. Using the modern notation l^2 to denote the square of side l, we have $l^2 = 1^2 + 1^2 = 2$.

Now (again using modern notation), suppose that *l* is rational, in which case we can suppose that $l = m/n$, where *m* and *n* are whole numbers. We can also suppose that *m* and *n* have no common divisor except 1, since any other common divisor could be divided out of *m* and *n* in advance, without changing *l*. Under these conditions we can derive a contradiction by the following series of implications (these probably go back to the Pythagoreans, but the first known hint of such a proof is in Aristotle's *Prior Analytics* 1.23):

$$l = m/n \Rightarrow 2 = m^2/n^2 \quad \text{(squaring both sides)}$$
$$\Rightarrow 2n^2 = m^2 \quad \text{(multiplying both sides by } n^2\text{)}$$
$$\Rightarrow m^2 \text{ is even}$$
$$\Rightarrow m \text{ is even, say, } = 2p \text{ (since the square of an odd number is odd)}$$
$$\Rightarrow 2n^2 = (2p)^2 = 4p^2 \quad \text{(substituting } m = 2p \text{ in } 2n^2 = m^2\text{)}$$
$$\Rightarrow n^2 = 2p^2$$
$$\Rightarrow n^2 \text{ is even}$$
$$\Rightarrow n \text{ is even}$$
$$\Rightarrow 2 \text{ divides both } m \text{ and } n,$$

contrary to the assumption of no common divisor.

Since it is contradictory to assume that *l* is a ratio of whole numbers, *l* is an irrational length. The Greeks often expressed this by saying that the side and hypotenuse of the right-angled triangle with equal sides are *incommensurable* – not whole number multiples of any common unit of measure.

In the view of the Pythagoreans, the lack of a common unit of length meant that lengths are not numbers, because 'numbers' to them were whole numbers and their ratios. In particular, sums and products of lengths are not necessarily like sums and products of numbers, so the concept of 'sum of squares' needs clarification. In the next section we will see how Euclid handled sums and products of lengths.

1.3 Operations on Lengths and Numbers

By denying that irrational lengths could be numbers, yet allowing that they could be squared and added, the Greek mathematicians after Pythagoras had to define sum and product in purely geometric terms.

The sum of two lengths is defined in the obvious way suggested by Figure 3. The lengths are represented by two line segments *a* and *b*, and $a + b$ is obtained by joining these segments end to end.

It follows easily that $a + b = b + a$ and $a + (b + c) = (a + b) + c$ (commutative and associative laws). It is also clear, since the sum of lengths is a length,

$$\frac{}{a} + \frac{}{b} = \frac{}{a+b}$$

Figure 3 The sum of two lengths.

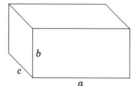

Figure 4 The product of three lengths.

that any number of lengths can be added. Thus lengths behave exactly like numbers as far as addition is concerned.

The behaviour of products is not so simple. The product of lengths a and b is not a length but the rectangle with perpendicular sides a and b. And the product of lengths a, b, and c is the rectangular box with perpendicular sides a, b, and c (Figure 4).

It is clear from these definitions that $ab = ba$ and $a(bc) = (ab)c$, and it can also be seen that $a(b+c) = ab + ac$ (the latter is actually a special case of Euclid's Proposition 1 of Book II of the *Elements*). Thus, to the extent that sum and product are defined, lengths satisfy the same laws as positive numbers. The trouble is that they are defined only to a limited extent, so the algebra of lengths is crippled. Products of more than three lengths are not admitted, because they have no geometric counterpart. Likewise, products can be added only when each is of the same 'dimension', that is, a product of the same number of lengths.

Finally, there is a complicated, though geometrically natural, notion of *equality*. It says, for example, that two rectangles R and S are equal if R can be cut into a finite number of triangles which reassemble to form S. We say more about Euclid's theory of equality for rectangles, and other polygons, in the next section. Remarkably, this theory is perfectly adequate for polygons, because any two polygons of equal area (in the modern sense) are actually equal in Euclid's sense. However, the theory is not adequate for polyhedra, as was shown by Dehn (1900). Dehn showed that a cube and regular tetrahedron of equal volume are not equal in Euclid's sense.

1.4 Axiomatics

The power of the axiomatic method is charmingly described by John Aubrey in his *Brief Lives*, speaking of Thomas Hobbes:

Figure 5 Non-parallel lines.

> He was 40 yeares old before he looked on geometry; which happened
> accidently. Being in a gentleman's library ... Euclid's *Elements* lay open,
> and 'twas the 47 *Elements*, Book I. He read the proposition. 'By G–', sayd he,
> 'this is impossible!' So he reads the demonstration of it, which referred him
> back to such a proposition; which proposition he read. That referred him back
> to another, which he also read ... that at last he was demonstratively con-
> vinced of that trueth. This made him in love with geometry.

Proposition 47 of Book I, incidentally, is the Pythagorean theorem. The
Elements is the first systematic account of theorems and proofs that has come
down to us, and it became the standard way of presenting mathematics in the
Western world (and later the Islamic world) for the next 2000 years.

Euclid begins with a small number of basic assumptions (axioms) and
deduces all theorems from them by logic. His axioms include simple statements about points, lines, length, and angle. There are also statements about
equality, addition, and subtraction, such as 'things which are equal to the
same thing are equal to each other' and 'if equals be added to equals then the
wholes are equal'. The principles of logic are not explicitly stated. The most
important axiom, needed for the Pythagorean theorem and many others, is
the parallel axiom. Euclid states it as follows, in the translation by Heath
(1956):

> That, if a straight line falling on two straight lines make the interior angles on the
> same side less than two right angles, the two straight lines, if produced indefi-
> nitely, meet on that side on which are the angles less than two right angles.

This rather long-winded statement is illustrated in Figure 5. The line n falls on
the lines l and m, making angles α and β on the right with $\alpha + \beta < \pi$. The
conclusion is that l and m then meet somewhere on the right. Thus the parallel
axiom actually gives a condition for lines not to be parallel.

It follows (not quite obviously) that there is exactly one parallel to a given
line l through a given point P outside l, namely the line m for which $\alpha + \beta = \pi$.

The complicated character of the parallel axiom provoked many attempts to
eliminate it by showing that it follows from Euclid's other axioms. But all such
attempts failed. This led, in the nineteenth century, to a thorough examination of the

axiomatic method and to subsequent analysis of its scope and limits. We pick up this story later.

1.5 Philosophical Issues

According to legend, the Pythagoreans were the first to propose a philosophy of mathematics, in fact a very simple 'theory of everything': *all is number.* It is said that they observed the role of whole numbers in musical harmony and jumped from there to the conclusion that the whole universe is ruled by whole numbers and their ratios. The echoes of this philosophy are still heard in phrases like 'the harmony of the spheres'.

Whatever its details may have been, the Pythagorean philosophy was disrupted by the discovery of irrational quantities such as $\sqrt{2}$. Irrationals were unacceptable as numbers, but unavoidable in geometry, since no one could deny that if a square exists, then so does its diagonal. This led to the separation of number theory and geometry seen in Euclid's *Elements* but also to the theory found in the *Elements* Book V. The 'theory of proportions' found in Book V establishes a point of contact between (rational) numbers and geometric quantities, though without fully reconciling the two.

Much of the subsequent history of mathematics, and its philosophy, grows from the struggle to reconcile the concepts of number and quantity, or the discrete and the continuous, or the rational (logical) and the visual (intuitive). The development of mathematical philosophy accompanies this struggle, as we will see in the sections that follow. At the end of each section I will give a historical update, as it were, of philosophical developments, under the headings of logic and intuition, meaning and existence, and discrete and continuous. As a mathematician, I prefer to think in these terms, but I hope that philosophers will be able to translate the philosophical content of my remarks into their own preferred terms.

Intuition and logic. Probably in an attempt to work precisely with geometric quantities, Euclid's *Elements* is the first known example of the axiomatic approach to truth, whereby theorems are deduced from axioms by logic. However, his axioms are incomplete, and there are frequent appeals to intuition, one even in his first proposition. Thus the *Elements* unintentionally illustrates how hard it is to avoid unconscious assumptions in mathematical reasoning.

Meaning and existence. Euclid also undercuts what we now consider to be the axiomatic method by attempting to define primitive concepts such as 'point' and 'line'. He also restricts the concept of 'number' essentially to the natural numbers and their ratios. The irrationality of $\sqrt{2}$ is thought to disqualify it from being a number, but Euclid did not prescribe what the properties of numbers should be.

Discrete and continuous. Because of its sensitivity to irrational quantities, the *Elements* generally makes a clear separation between the concepts of number and quantity, or between the discrete and the continuous. But Book V begins a possible merger between the two, as we will see later. This illustrates the sometimes opposing tendencies of mathematics and philosophy. Mathematicians generally have the outlook expressed by Poincaré in 1908:

> I think I have already said somewhere that mathematics is the art of giving the same name to different things. It is enough that these things, though differing in matter, should be similar in form, to permit of their being, so to speak, run in the same mould. When language has been well chosen, one is astonished to find that all demonstrations made for a known object apply immediately to many new objects: nothing requires to be changed, not even the terms, since the names have become the same. (see Poincaré 1952, 34)

In other words, mathematicians consider things to be the same if they have the same behaviour. Philosophers, however, like to make distinctions: they look for reasons why things should *not* be considered the same. Sometimes a distinction seems to be justified, as was the Greek distinction between numbers and geometric quantities such as length. But mathematics seeks to erase distinctions where possible. The long evolution of the real number concept can be viewed as a project to erase the distinction between number and quantity and, with it, the distinction between real number theory and geometry.

2 Infinity in Greek Mathematics

PREVIEW

Although number and length are mostly kept separate in the *Elements*, there is one process that Euclid applied to both – the Euclidean algorithm, which operates on a pair by 'repeatedly subtracting the smaller from the larger'. When applied to a pair of positive integer numbers (or more generally, to a pair of positive integer multiples of a unit length) the algorithm terminates because positive integers cannot decrease forever. But when applied to a pair of lengths in irrational ratio, the algorithm does not terminate. Indeed, Euclid used non-termination of his algorithm as a criterion for irrationality, thus bringing infinity into the discussion of irrationality.

The number-free theory of area, used by Euclid to prove the Pythagorean theorem, works quite smoothly for areas of polygons. But a similar approach to volume fails for even simple polyhedra, such as the tetrahedron. Euclid was able to find the volume of the tetrahedron by decomposing it into infinitely many prisms, thus bringing infinity into the theory of volume. The Greek theory of area also had difficulty with curved regions, which obviously cannot be decomposed into finitely many polygons. However, Archimedes was able to find the area of a parabolic segment by decomposing it into infinitely many triangles. Nevertheless, the Greeks sought to 'avoid infinity' by considering arbitrary finite sums instead of infinite sums.

2.1 Irrationality and Non-termination

The Euclidean algorithm is introduced in Book VII of the *Elements*, as a method for finding the greatest common divisor of two positive integers. As Euclid says, one 'continually subtracts the less from the greater'; more precisely, if $a > b$, one replaces the pair a,b by $a - b,b$. Since positive integers cannot decrease forever, the algorithm always terminates. For example, with the pair 13, 8, one gets

$$13, 8 \rightarrow 5, 8 \rightarrow 5, 3 \rightarrow 2, 3 \rightarrow 2, 1 \rightarrow 1, 1.$$

When a pair of identical numbers is obtained, that number is $\gcd(a,b)$, because all common divisors of the pair are preserved by subtraction. Thus our example shows that $\gcd(13, 8) = 1$.

In Book X of the *Elements* Euclid generalizes the algorithm to lengths a and b, in which case it may not terminate. For example, if (using modern notation) the lengths are $a = \sqrt{2}$ and $b = 1$, then the first two steps are

$$\sqrt{2}, 1 \rightarrow 1, \sqrt{2} - 1 \rightarrow 2 - \sqrt{2}, \sqrt{2} - 1.$$

At this point it may be noticed that $2 - \sqrt{2}$ and $\sqrt{2} - 1$ are in the same ratio as $\sqrt{2}$ and 1. It is not clear whether the Greeks noticed this (though they were probably aware of something similar), but it is clear by basic algebra because

$$2 - \sqrt{2} = \sqrt{2}(\sqrt{2} - 1).$$

Since $2 - \sqrt{2}$ and $\sqrt{2} - 1$ are in the same ratio as $\sqrt{2}$ and 1, applying the Euclidean algorithm to them will produce, in two steps, yet another pair in that ratio – and so on, forever.

Whether or not Euclid knew this particular example, he realized that the algorithm does not terminate on a pair of lengths in irrational ratio (Book X,

Figure 6 Equality of parallelogram and rectangle.

Figure 7 Equality of triangle and half rectangle.

Proposition 2). Thus the Euclidean algorithm elegantly separates rational from irrational, by separating termination from non-termination; that is, finite from infinite.

2.2 Areas and Volumes

In Book I of the *Elements* Euclid shows equality of various regions by adding or subtracting equal triangles. For example, Figure 6 shows that a parallelogram equals a rectangle of the same base and height. And Figure 7 shows that a triangle equals half a rectangle with the same base and height.

There are also decompositions showing that any rectangle equals a rectangle with a given base. Using this fact, it is possible to find a rectangle equal to any polygon, by cutting the polygon into finitely many triangles. Now if (as in the case of the triangle) one region R equals a rational multiple r of some standard region we can take as a unit, then it is compatible with Euclid to let the number r 'measure' the region R in the same way that we measure the area of R. Under these conditions we will speak of numerical 'areas' and 'volumes' from now on.

Now a curved region obviously cannot be cut into finitely many triangles. The best we can hope for is a decomposition into infinitely many triangles which, if we are lucky, might be comprehensible. Archimedes had a brilliant success by this method, finding the area of a parabolic segment.

The parabolic segment is filled with triangles in the manner shown in Figure 8: first the black triangle, then two dark grey triangles below it, then four lighter grey triangles below them, and so on.

Each triangle is half the width of the triangle above it, and a calculation shows that each group (of one, two, four, … triangles) has total area one-fourth the area

Figure 8 Filling the parabolic segment with triangles.

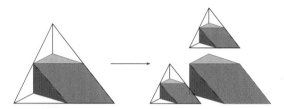

Figure 9 Filling the tetrahedron with prisms.

of the group above it. Thus, if the black triangle has area 1, then the total area of the triangles is the infinite sum

$$1 + \frac{1}{4} + \left(\frac{1}{4}\right)^2 + \left(\frac{1}{4}\right)^3 + \left(\frac{1}{4}\right)^4 + \ldots.$$

If we set

$$S = 1 + \frac{1}{4} + \left(\frac{1}{4}\right)^2 + \left(\frac{1}{4}\right)^3 + \left(\frac{1}{4}\right)^4 + \ldots,$$

then clearly

$$4S = 4 + 1 + \frac{1}{4} + \left(\frac{1}{4}\right)^2 + \left(\frac{1}{4}\right)^3 + \left(\frac{1}{4}\right)^4 + \ldots,$$

so $4S - S = 4$ and therefore $S = 4/3$. (This risky calculation with infinite sums gives a quick way to guess the answer. We will see how Archimedes did it more carefully in the next section.)

The tetrahedron is filled with prisms, as indicated in Figure 9. After two prisms are removed (one light grey, the other darker grey), two smaller tetrahedra remain, from which we again remove prisms, and continue.

In this way Euclid found that the volume of the tetrahedron is a similar infinite sum,

$$\frac{1}{4} + \left(\frac{1}{4}\right)^2 + \left(\frac{1}{4}\right)^3 + \left(\frac{1}{4}\right)^4 + \ldots,$$

which equals 1/3.

2.3 The Method of Exhaustion

The last two sections have shown examples of infinite processes in Greek mathematics. Those in the previous section are notable for finding a rational area or volume apparently not obtainable by finite processes. Nevertheless, the Greeks were suspicious of infinity and they tried to avoid it as far as possible. In many cases they were able to do so by what is called the *method of exhaustion*.

The method is so called because one confirms a result x, obtained by an infinite process, by exhausting all other possibilities (less than x or greater than x). Typically, the other possibilities are exhausted by supposing the process to run for an arbitrary, but finite, number of steps. We can illustrate the method of exhaustion in the case of Archimedes' evaluation of the area of a parabolic segment, replacing the infinite processes by arbitrary finite ones. The infinite processes involved in this example are as follows:

1   Filling the parabolic segment with infinitely many triangles. We can argue that the triangles do fill the segment (or 'exhaust' it) by showing that any given point inside the parabola falls inside some triangle. This is a purely finite, if tedious, argument.
2   Showing that the infinite series

$$1 + \frac{1}{4} + \left(\frac{1}{4}\right)^2 + \left(\frac{1}{4}\right)^3 + \left(\frac{1}{4}\right)^4 + \ldots \text{ has sum } 4/3.$$

This can be done by finding the finite sum

$$S_n = 1 + \frac{1}{4} + \left(\frac{1}{4}\right)^2 + \ldots + \left(\frac{1}{4}\right)^n = \frac{4}{3} - \frac{1}{3}\left(\frac{1}{4}\right)^n.$$

We see that S_n 'exhausts' all numbers $< 4/3$ by taking arbitrarily large values of n, because $\left(\frac{1}{4}\right)^n$ then becomes arbitrarily small. And obviously S_n cannot be $> 4/3$, so the value that remains, 4/3, is necessarily the sum of the infinite series.

Thus the area of the parabolic segment (and similarly the volume of the tetrahedron) can be found by using arbitrary finite sums instead of infinite sums. This is typical of the way the Greeks avoided actual use of infinity. We will see a similar 'avoidance of infinity' in the next section.

2.4 The Theory of Proportions

A common stumbling block for readers of the *Elements* was Book V, on the so-called theory of proportions. This theory, thought to be due to Eudoxus, is about the concept of length and its relation to the concept of number. Its main difficulty is basically the difficulty we meet today in trying to understand the real number line. There is a lesser difficulty due to the Greek habit of working with ratios (of integers or lengths), which stopped them from working with rational numbers and a unit length. But I will gloss over this in order to concentrate on the main difficulty – which is of course the existence of irrationals – by admitting rational numbers and a unit of length.

When rational numbers are admitted it becomes clear that they are the key ingredient in the theory of proportions. Given a unit length 1, we also have a rational length $l = m/n$ for any positive integers m, n. It is the length such that n copies of l equal m copies of 1. These rational lengths determine, for any lengths a and b, whether $a < b$, $a = b$, or $a > b$. Namely

$a < b \Leftrightarrow$ there is a rational m/n with $a < m/n < b$;

$a > b \Leftrightarrow$ there is a rational m/n with $a > m/n > b$;

$a = b \Leftrightarrow$ neither $a < b$ nor $a > b$.

Thus if $a < b$ or $a > b$, there is a pair m, n that 'witnesses' the fact: either because $a < m/n < b$ or $a > m/n > b$. But if $a = b$, there is no single rational number m/n that witnesses this fact (unless a and b are rational): a equals b only if all rational numbers less than a are less than b, and conversely. Thus equality is more elusive than inequality, and irrational lengths are more elusive than rational lengths.

Because of their suspicion of infinity, the Greeks did not take the step of saying that an irrational length is determined by the rational lengths on either side of it, since this determination involves infinitely many rational lengths. In Section 6 we will see what happened when this step was taken, in the nineteenth century.

Even the test for inequality raises an interesting philosophical point. Suppose that $0 < b$ and hence that $0 < m/n < b$ for some positive integers m and n. It follows, multiplying by n, that $0 < m < nb$. Thus the theory of proportions assumes what would later be called the *Archimedean axiom* or non-existence of infinitesimals: if $b > 0$ then some integer multiple $nb > 1$. This became a hot issue in the seventeenth century, when mathematicians found it convenient to assume the existence of infinitesimals; that is, they supposed there were $b > 0$ such that no integer multiple $nb > 1$ (see Section 4.3).

2.5 Archimedes and Actual Infinity

Like the method of exhaustion, the theory of proportions avoids infinity by dealing instead with the 'arbitrary finite'. In particular, it employs the arbitrary finite integers m, n to witness that $a < b$ because $a < m/n < b$. Indeed, when $a = b$ is proved it is done by exhaustion; namely, by showing that $a < m/n < b$ and $a > m/n > b$ fail for any possible pair m, n.

An infinity that can be 'exhausted' in this way by finite parts was called a 'potential infinity' by the Greeks, and only potential infinities were allowed in mathematical argument. The aim of the method of exhaustion was to avoid *actual* infinity; that is, viewing an infinite collection or process as a completed whole. For over 2000 years after Euclid the official practice of mathematicians was to accept only potential infinities, and to avoid actual infinities by appealing to the method of exhaustion. This was the theory, but in practice actual infinities were often used as a shortcut to results that (hopefully) could later be proved rigorously by the method of exhaustion.

As far as we know, the first to use actual infinities was Archimedes, in a work called *The Method*. This work was lost and unknown for centuries before being rediscovered in 1906, so it did not influence the development of mathematics or its philosophy. However, it does show that there was more to Greek thinking about infinity than one would gather from Euclid. Archimedes went far beyond thinking that a sequence of discrete steps could be completed; he was even willing to view the continuum of points on the line as a completed collection (though probably without realizing that the continuum is a new kind of infinity).

2.6 Philosophical Issues

Euclid penetrated to the heart of the distinction between rational and irrational: it is the difference between finite and infinite. Thus geometry, which accepts irrational quantities, must also accept infinity in some form. So the logic of geometry had to find a way of reasoning about infinity.

Intuition and logic. The Greeks tried as far as possible to avoid reasoning about infinity. The origin of their fear of infinity is not completely clear, though Zeno's paradoxes show that the fear was present before Euclid's time, and that Aristotle tried to debunk it. The method of exhaustion was the mathematical response to these philosophical debates: avoid reasoning about infinity by reasoning about finite (but arbitrary) stages of an infinite process, and argue that they 'exhaust all possibilities' except one – which possibility can therefore be deemed the result of the infinite process.

Meaning and existence. Thus the method of exhaustion indirectly gives meaning to the result of an infinite process, without commitment to the existence of actual infinity.

Continuous and discrete. Processes with an infinite number of discrete steps could be accepted as 'potential infinity'. But, in rejecting actual infinity, the Greeks generally stayed well away from the idea of a 'continuous infinity'. Perhaps Archimedes was an exception, because he seemed willing to accept the continuum of points on the line as a completed whole.

As mentioned in Section 1.5, the tendency of philosophy is to make distinctions, while that of mathematics is to erase them where possible. However, mathematics sometimes discovers distinctions that philosophy (and previous mathematics) had not foreseen. Infinity is a case in point. For millennia, mathematics accepted the distinction between potential and actual infinity and, following philosophy, considered actual infinity unacceptable. But since the late nineteenth century, actual infinity has not only become acceptable (to most mathematicians) but also subject to unforeseen and complicated distinctions. This has led to new controversies about where to draw the line between acceptable and unacceptable infinity. We will see how this happened in Section 7.

3 Imaginary Numbers

PREVIEW

Resistance to treating quantities such as $\sqrt{2}$ as numbers gradually eroded over the centuries, possibly because of the rise of algebra in India and the Islamic world. In India, Brahmagupta around 600 CE gave essentially the modern solution $x = \frac{-b \pm \sqrt{b^2 - 4ac}}{2a}$ of the quadratic equation $ax^2 + bx + c = 0$, and it became accepted that the square root of a positive number was itself a number. At the same time, there was reluctance to use negative numbers (though Brahmagupta accepted them), and the square root of a negative number seemed nonsensical. Thus when $b^2 - 4ac < 0$ it seemed natural to say that $ax^2 + bx + c = 0$ had no solution.

Things changed in the sixteenth century when Italian mathematicians discovered the solution of the cubic equation $x^3 = px + q$ in the form

$$x = \sqrt[3]{\frac{q}{2} + \sqrt{\left(\frac{q}{2}\right)^2 - \left(\frac{p}{3}\right)^3}} + \sqrt[3]{\frac{q}{2} - \sqrt{\left(\frac{q}{2}\right)^2 - \left(\frac{p}{3}\right)^3}}.$$

This formula called not only for acceptance of square roots and cube roots, but also for square roots of negative numbers – because there are equations for which there is an obvious real solution yet $\left(\frac{q}{2}\right)^2 - \left(\frac{p}{3}\right)^3 < 0$.

For a long time, numbers such as $\sqrt{-1}$ were called *impossible*, and they are still called 'imaginary'. Yet they were accepted in mathematics, at least to prove results about real numbers, because they were useful and they did not (usually) lead to contradiction. Eventually, the system of real and imaginary numbers came to be viewed as natural, both algebraically and geometrically.

3.1 Quadratic and Cubic Equations

As we have seen in Section 1, algebra was hamstrung in Greek mathematics by the geometric interpretation of product. Under the geometric interpretation, products of more than three terms have no meaning, and products of different dimensions cannot be added. These restrictions do not apply to the product of numbers, so in a contest between algebra for numbers and algebra for lengths, algebra for numbers clearly wins.

This is roughly what happened in the development of algebra, which was initially a symbolism for solving problems (especially equations) about numbers. Until about 1600, algebra was a discipline that solved equations in numbers but it fell back on geometry to justify its moves – because Euclid's *Elements* was still the model for mathematical proof.

The shift from geometry to an algebra of numbers began with Diophantus, around 200 CE, in the last phase of classical Greek mathematics. Diophantus used a symbolism that allowed products of four or more elements. But because he was interested in finding rational solutions of equations he used only the rational operations $+, -, \times, \div$, not the square root operation $\sqrt{\ }$.

The $\sqrt{\ }$ operation occurs in the general solution

$$x = \frac{-b \pm \sqrt{b^2 - 4ac}}{2a} \tag{*}$$

of the quadratic equation $ax^2 + bx + c = 0$. Essentially, this solution was given by Brahmagupta in India around 600 CE, though in words rather than symbols. The study of equations spread from India to the Islamic world, where it was given the name 'algebra' by al-Khwarizmi around 800 CE. From there it passed to Italy, where the next major advance occurred: the solution of cubic equations.

The solution of $x^3 = px + q$ was discovered by Scipione del Ferro around 1500 but was kept as a 'secret weapon' for the mathematical contests that were then popular. It was rediscovered by Tartaglia in the 1530s and first published in the *Ars Magna* of Cardano (1545). The so-called Cardano formula for the solution is

$$x = \sqrt[3]{\frac{q}{2} + \sqrt{\left(\frac{q}{2}\right)^2 - \left(\frac{p}{3}\right)^3}} + \sqrt[3]{\frac{q}{2} - \sqrt{\left(\frac{q}{2}\right)^2 - \left(\frac{p}{3}\right)^3}}. \qquad (**)$$

Mathematicians by this time were willing to accept square and cube roots of positive numbers, but they balked at square roots of negative numbers. Of course square roots of negative numbers already occur in the quadratic formula (*) when $b^2 < 4ac$. But in this case one is free to say that the equation $ax^2 + bx + c = 0$ has no solution.

It was otherwise with the solution (**) of the cubic equation.

3.2 Bombelli's Algebra of Imaginary Numbers

The equation $x^3 = px + q$ can have an obvious solution, even though the Cardano formula contains the square root of a negative number $\left(\frac{q}{2}\right)^2 - \left(\frac{p}{3}\right)^3$. This is the case, for example, for $x^3 = 15x + 4$, which has the obvious solution $x = 4$.

For this equation the Cardano formula gives (after some simplification)

$$x = \sqrt[3]{2 + 11\sqrt{-1}} + \sqrt[3]{2 - 11\sqrt{-1}}.$$

To reconcile this expression with the value $x = 4$, Bombelli (1572) assumed that $\sqrt{-1}$ obeys the same algebraic rules as ordinary numbers. He kept his calculation a secret, but it is easy to reconstruct. Using the modern notation i for $\sqrt{-1}$, so $i^2 = -1$, one can check that

$$(2 + i)^3 = 2 + 11i \text{ and } (2 - i)^3 = 2 - 11i,$$

and therefore

$$\sqrt[3]{2 + 11\sqrt{-1}} + \sqrt[3]{2 - 11\sqrt{-1}} = \sqrt[3]{2 + 11i} + \sqrt[3]{2 - 11i}$$

$$= \sqrt[3]{(2 + i)^3} + \sqrt[3]{(2 - i)^3}$$

$$= (2 + i) + (2 - i)$$

$$= 4.$$

As many people have since remarked, it seems as though algebra is smarter than we are!

At any rate, Bombelli's example, and others like it, eventually convinced mathematicians that it was safe to use 'imaginary' or 'impossible' numbers. Whatever $\sqrt{-1}$ means, if anything, it seems to behave like an ordinary number and to give correct results about ordinary numbers.

3.3 The Convenience of Imaginary Numbers

In the eighteenth and nineteenth centuries mathematicians discovered many situations in which known properties of ordinary numbers are more easily stated, or explained, with the help of imaginary numbers. Here are some examples.

1 The trigonometric formulas

$$\cos(\theta + \varphi) = \cos\theta \cos\varphi - \sin\theta \sin\varphi$$

$$\sin(\theta + \varphi) = \sin\theta \cos\varphi + \cos\theta \sin\varphi$$

are more concisely (and memorably) expressed by the single formula

$$\cos(\theta + \varphi) + i\sin(\theta + \varphi) = (\cos\theta + i\sin\theta)(\cos\varphi + i\sin\varphi).$$

2 The latter formula can be even more concisely expressed by

$$e^{i(\theta+\varphi)} = e^{i\theta}e^{i\varphi},$$

if we assume $e^{ix} = \cos x + i\sin x$ (see Section 4.5 for a reason to do this). This gives new meaning to the sine and cosine functions – as parts of the imaginary exponential function.

3 If a,b,c,d are positive integers then the product of $a^2 + b^2$ and $c^2 + d^2$ is itself the sum of two integer squares. To find the latter squares we use $i^2 = -1$ to create the imaginary factorizations

$$a^2 + b^2 = a^2 - (ib)^2 = (a + ib)(a - ib)$$

$$c^2 + d^2 = c^2 - (id)^2 = (c + id)(c - id),$$

using the identity $x^2 - y^2 = (x + y)(x - y)$. Combining the factorizations, and again using the identity, we get

$$(a^2 + b^2)(c^2 + d^2) = [(a + ib)(c + id)][(a - ib)(c - id)]$$
$$= [ac - bd + i(ad + bc)][ac - bd - i(ad + bc)]$$
$$= (ac - bd)^2 + (ad + bc)^2.$$

These examples suggest that imaginary numbers should be accepted, if only for the sake of convenience. However, it is possible to do better than this. We can give a convincing interpretation of imaginary numbers, which shows them to be just as 'real' as ordinary numbers (and incidentally explains their role in geometry and trigonometry). More conservatively, one can show how to eliminate imaginary numbers, from any argument that uses them, in favour of ordinary numbers.

3.4 Realizing the Imaginary

A simple way to eliminate imaginary numbers from mathematics was introduced by Hamilton (1835): replace $a + ib$ by the ordered pair (a,b) of real numbers a,b, and define the sum and product of pairs by

$$(a_1,b_1) + (a_2,b_2) = (a_1 + a_2, b_1 + b_2)$$
$$(a_1,b_1) \cdot (a_2,b_2) = (a_1 a_2 - b_1 b_2, a_1 b_2 + a_2 b_1).$$

Then the pair for the sum $(a_1 + ib_1) + (a_2 + ib_2)$ is the sum of the pairs for $a_1 + ib_1$ and $a_2 + ib_2$, and the pair for the product is likewise the product of the pairs. It follows that any statement about sum and product of numbers of the form $a + ib$ is equivalent to one about real numbers, for example

$$(a_1 + ib_1)(a_2 + ib_2) = c + id \iff a_1 a_2 - b_1 b_2 = c \text{ and } a_1 b_2 + a_2 b_1 = d.$$

Hence any argument involving i can be replaced by one involving real numbers alone.

Because of this we say that the theory of complex numbers (as the numbers $a + ib$ are called) is a *conservative extension* of the theory of real numbers. It is 'conservative' in the sense that any result about real numbers proved with use of i can be proved without it. Hamilton's construction shows that it is harmless to assume that imaginary numbers exist, but at the same time it shows that there is no need to assume they exist. Anything we can do with them we can do without them, though perhaps not as easily.

For most mathematicians, what compels belief in the complex numbers is that they give more than we asked for. It is as though they were always part of the fabric of mathematics, but at first we noticed only one small thread in the solution of cubic equations. In fact, i not only gives solutions to cubic equations but to all polynomial equations. This is the fundamental theorem of algebra that we will say more about in Section 5.

Moreover, while i cannot lie on the line of real numbers, it makes perfect sense for it to lie on a perpendicular line of imaginary numbers. All we have to

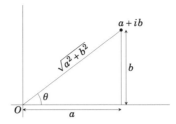

Figure 10 The geometry of a complex number.

do is imagine a plane of complex numbers, with $a + ib$ represented by the point (a,b) at horizontal distance a from the origin O, and at vertical distance b. Under this interpretation, multiplication by $a + ib$ has a geometric meaning that is a sweeping generalization of $i^2 = -1$. Namely multiplication by $a + ib$ magnifies the plane by $\sqrt{a^2 + b^2}$ (the distance of $a + ib$ from O) and rotates it about O through angle $\theta = \tan^{-1} \frac{b}{a}$. Figure 10 shows $a + ib$ in its geometric context.

In particular, multiplication by i rotates the plane about O through a right angle, and multiplication by $\cos \theta + i \sin \theta$ rotates it through angle θ. Thus multiplication of complex numbers encapsulates all of basic geometry and trigonometry. And this is just the beginning. As mentioned in the previous section,

$$e^{i\theta} = \cos \theta + i \sin \theta,$$

so the complex numbers unite trigonometry with the exponential function. In short, the complex numbers are probably the most powerful unifying and simplifying force in higher mathematics. That is why mathematicians believe in them.

3.5 Philosophical Issues

The emergence of algebra in India and the Islamic world did not at first affect the philosophy of mathematics. In fact, Euclid's authority ruled algebra until the sixteenth century. The Islamic algebraists justified their proofs by appeal to geometric diagrams, and such proofs occur as late as Cardano (1545). It was only around the end of the sixteenth century, when modern algebraic symbolism developed, that algebraic calculation became a new force in mathematics. Symbolic algebra paved the way for other branches of symbolic mathematics, such as calculus, which the philosophy of mathematics eventually had to take into account. Nevertheless, for a long time it was hoped that Euclid could remain as the foundation of mathematics.

Intuition and logic. The Italian algebraists at first justified the rules of algebra (as did their Islamic predecessors) by appeal to geometric logic. But Bombelli's calculations with $\sqrt{-1}$ suggested that algebra had an independent logic of its own.

Meaning and existence. Imaginary numbers at first were accorded merely 'symbolic existence', allowing them to be used in calculations that had real results. In becoming a system of rules for manipulating symbols, algebra began to break free from its geometric foundation. Yet, surprisingly, imaginary numbers provide new insights into geometry.

Discrete and continuous. The Greek belief that only rational numbers were really numbers gradually eroded under the influence of algebra, which urged acceptance of square and cube roots, and of trigonometry, which urged acceptance of the sine and cosine functions. However, there was not yet a coherent theory of real numbers – only the belief that they could be modelled by the points of a line.

4 Calculus and Infinitesimals

Preview

The influence of algebra grew in the seventeenth century, first in the algebraic geometry of Fermat and Descartes, then in the infinitesimal calculus of Newton and Leibniz. In geometry, algebra made quick work of ancient problems about lines and conic sections and gave easy access to a vast class of curves barely touched by the Greeks. The algebraic approach to geometry was made possible by *arithmetization* of the line and plane: identifying points of the line with real numbers and points of the plane with pairs of real numbers. Under this identification, many curves could be described by polynomial equations, $p(x,y) = 0$, allowing geometric properties to be extracted by algebraic manipulation.

Calculus extended this idea by allowing algebraic operations on *infinitesimals* – quantities that behaved like non-zero numbers in calculations but were otherwise negligible. For example, the slope of a curve $p(x,y) = 0$ could be calculated as a quotient dy/dx of infinitesimals dx and dy, where dx was taken to be an infinitesimal increase in x and dy the corresponding increase in y.

The properties ascribed to infinitesimals were close to, if not actually, inconsistent. Yet, like imaginary numbers, infinitesimals seemed easy and safe to use. Mathematicians believed that, if challenged, they could reproduce the results of infinitesimal calculus by the more rigorous method of exhaustion. The magic of infinitesimal calculus was its ability to replace complicated exhaustion arguments by routine calculations, so once again convenience overcame any doubts about the existence of the mathematical objects being used.

4.1 Infinite Series

We have now seen that Euclid and Archimedes used sums of infinite series to find certain areas and volumes. The series in question were instances of the infinite geometric series $a + ar + ar^2 + ar^3 + \ldots$, which has sum $\frac{a}{1-r}$ when $|r| < 1$. This value can be rigorously confirmed by the method of exhaustion. We find that the finite series

$$a + ar + ar^2 + ar^3 + \ldots + ar^n = \frac{a - ar^{n+1}}{1 - r},$$

and this finite sum (for $a > 0$ and $|r| < 1$) is clearly less than $\frac{a}{1-r}$ but able to exceed any number less than $\frac{a}{1-r}$, since r^{n+1} can be made arbitrarily small by choosing n sufficiently large.

Therefore, the finite sums 'exhaust' all numbers less than $\frac{a}{1-r}$ and so the infinite sum must equal $\frac{a}{1-r}$.

When calculus was invented, around 1665, the geometric series was the starting point for many other results on infinite series. However, before calculus was invented, remarkable results about infinite series in trigonometry were discovered in fifteenth-century India. The main contributor to these discoveries was Madhava (c. 1340–c. 1425) and his methods were largely algebraic. The starting point was again the geometric series, but new series were also used ingeniously, notably the series

$$1^k + 2^k + 3^k + \ldots + n^k \text{ for } k = 1, 2, 3, \ldots .$$

The latter series played a role later taken over by calculus in proving that

$$\tan^{-1} x = x - \frac{x^3}{3} + \frac{x^5}{5} - \frac{x^7}{7} + \ldots \text{ for } -1 < x \leq 1.$$

Madhava also discovered the series for the sine and cosine functions:

$$\sin x = x - \frac{x^3}{3!} + \frac{x^5}{5!} - \frac{x^7}{7!} + \ldots$$

$$\cos x = 1 - \frac{x^2}{2!} + \frac{x^4}{4!} - \frac{x^6}{6!} + \ldots .$$

The latter series, and the related series for e^x, were rediscovered in Europe in the seventeenth century, and they played an important role in the development of calculus. The independent discovery of these results in India and Europe was perhaps the most remarkable example of the cultural universality of mathematics since the Pythagorean theorem.

4.2 Algebraic Geometry

Infinite processes on numbers were one prerequisite for calculus. Another was the application of algebra to geometry, or *algebraic geometry* for short. The latter became possible after algebraic symbolism came to maturity in the sixteenth century, allowing calculations with polynomials to be made just as easily as with numbers.

By the 1630s, Fermat and Descartes were able to give an algebraic solution of a problem that is usually solved by calculus today: finding the tangents to an algebraic curve. The setup for this problem is one that is now familiar to high school students. Each point P in the plane is given by an ordered pair (x,y) of numbers, where

$x =$ horizontal distance to P from the origin O,

$y =$ vertical distance to P from the origin O.

An algebraic curve is one whose points satisfy an equation $p(x,y) = 0$, where p is a polynomial.

For example, the points at distance 1 from O satisfy $x^2 + y^2 = 1$ so the equation for the unit circle is $x^2 + y^2 - 1 = 0$. Another example is the parabola $y = x^2$, or $y - x^2 = 0$.

This leads to a classification of curves by the degree of the polynomial. If p is of degree 1 – that is $p(x,y) = ax + by + c$ – then $p(x,y) = 0$ is the equation of a line. If $p(x,y)$ is of degree 2, then Fermat and Descartes discovered (independently) that $p(x,y) = 0$ is one of the conic sections studied by the Greeks. Apart from degenerate cases, where the plane cutting the cone meets it in a point or lines, these are the ellipses, parabolas, and hyperbolas. Typical examples of these three types are shown in Figure 11.

Now a tangent to a curve k is a line l that meets k at a point 'multiply', in the sense that there is a multiple solution to the equation for x that results from

Figure 11 Ellipse, parabola, and hyperbola.

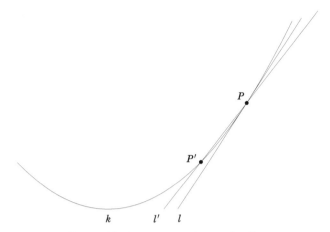

Figure 12 A tangent and a nearby line.

substituting the expression for y in l in the equation $p(x,y) = 0$. This precise algebraic condition captures the vague idea that l meets k at a single isolated point P, but that any line l' through P and close to l meets k in P and at least one other point P' near P. Figure 12 shows the situation. The 'double contact point' P of l is the limit of the two contact points P,P' of l', as the line l' approaches l.

For example, the line $y = 2x - 1$ is tangent to the parabola $y = x^2$ at x, because when we substitute $y = 2x - 1$ in $y = x^2$ we get the equation $2x - 1 = x^2$, or $x^2 - 2x + 1 = 0$. This equation can be rewritten $(x - 1)^2 = 0$, which shows that it has the double solution $x = 1$.

Similar calculations (though naturally more complicated if the curve has higher degree) allow us to find the tangents to any algebraic curve. However, finding the area between an algebraic curve and the x-axis is a more formidable problem, even for the curve $y = 1/x$. This is where the need for calculus becomes acute.

4.3 Infinitesimal Calculus

The algebraic criterion for tangency is simple to state, though it can be hard to use. A more serious objection is that it applies only to algebraic curves, and some physically natural curves are not algebraic. One famous example is the catenary, the shape of a hanging chain. The catenary looks rather like a parabola, but it is not, so another method is needed to find its tangents. In fact, even for algebraic curves a simpler method is desirable – one that finds the slope at any point. A better method is that of differential calculus, a system that arose from the algebra and geometry of hypothetical entities called infinitesimals.

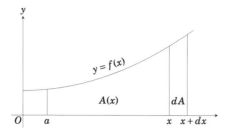

Figure 13 Infinitesimal geometry of area.

Infinitesimals were used by many of the early exponents of calculus, starting in the 1630s and coming to maturity with the infinitesimal calculus of Leibniz in the 1680s. Leibniz introduced the notations dx, dy, dz and the like for infinitesimals, which he seemed to view as quantities smaller than ordinary numbers, yet non-zero. For example, the slope of a curve at a point (x, y) was viewed as the slope dy/dx between (x, y) and a point $(x + dx, y + dy)$ on the curve 'infinitesimally close' to (x, y).

Given an equation $y = f(x)$ for the curve, it was generally easy to calculate the slope dy/dx. Take the parabola $y = x^2$ for example. In this case,

$$dy = (x + dx)^2 - x^2 = 2x \, dx + (dx)^2,$$

so

$$\frac{dy}{dx} = \frac{2x \, dx + (dx)^2}{dx} = 2x + dx.$$

At this stage one feels free to neglect dx and conclude that the slope of $y = x^2$ for any value of x is $2x$. (In particular, when $x = 1$ the slope is 2, so the equation of the tangent at this point is $y = 2x - 1$, as found by conventional algebra in the previous section.)

Similar calculations with dx and dy easily yield the slope of any algebraic curve, and hence the tangent, at any point on the curve. In particular, the slope of $y = x^n$ is nx^{n-1}. But this is just a small taste of the magic of infinitesimals. They also allow the calculation of curved areas, such as the area under a curve $y = f(x)$. To do this one views the area as a function $A(x)$ of x, taking the region between a fixed value a and a variable value x, as in Figure 13.

An infinitesimal increase dx in x produces an infinitesimal increase dA in area, which we can write

$$dA = f(x)dx,$$

since the extra strip dA of area has width dx and height that differs only infinitesimally from $f(x)$. (We are again choosing a convenient moment to neglect infinitesimals.) We conclude, dividing both sides by dx, that

$$\frac{dA(x)}{dx} = f(x).$$

In other words, $A(x)$ is a function whose graph has slope is $f(x)$. Thus finding areas under curves is the *inverse problem* to finding slopes.

If $f(x)$ is a function we have already found as a slope dy/dx, then we can conclude that the area function $A(x)$ is the same as y, at least within a constant. For example, if $A(x)$ is the area under the parabola $y = x^2$ between 0 and x, then

$$\frac{dA}{dx} = x^2,$$

and we may conclude $A(x) = \frac{1}{3}x^3$, because $\frac{dy}{dx} = x^2$ when $y = \frac{1}{3}x^3$, and the functions $A(x)$ and y agree when $x = 0$.

This, in a nutshell, is the *infinitesimal calculus* of Leibniz. The inverse relation between the area and tangent problems is called its fundamental theorem. Newton discovered a similar calculus, though without the convenient and suggestive dx notation. The art of infinitesimal calculus was basically the algebra and geometry of infinitesimals, allied with some good judgement about when to 'neglect' infinitesimals. (For example, divide by dx before neglecting it!)

4.4 Infinitesimals: Criticism and Avoidance

Infinitesimals were a bit like imaginary numbers. They seemed to contradict accepted principles – just as imaginaries contradicted the principle that squares are positive, infinitesimals contradicted the Archimedean axiom for geometric quantities, stated in Section 2.4 – yet they enabled calculations that were otherwise difficult or impossible. For mathematicians of the seventeenth and eighteenth centuries this was generally good enough reason to accept them.

But in another way infinitesimals were not like imaginaries. While it is true that imaginaries are not part of the ordinary number system, that system can easily be enlarged to accommodate them – for example, by defining imaginaries as ordered pairs of ordinary numbers, as Hamilton did in 1835. It is not nearly as easy, or convenient, to enlarge the number system to include infinitesimals. It was not even known to be possible until the twentieth century, long after mathematicians had decided that it was better to avoid infinitesimals the way Euclid and Archimedes avoided infinity.

When infinitesimal calculus was in its infancy it was fairly easy, though tedious, to replace infinitesimal arguments by the method of exhaustion. But as infinitesimal algebra and geometry grew in power, and faith in its correctness grew stronger, it became a thankless task to rewrite arguments in the rigorous ancient manner. As early as 1659, Huygens wrote,

> Mathematicians will never have enough time to read all the discoveries in Geometry (a quantity which is increasing from day to day and seems likely in this scientific age to develop to enormous proportions) if they continue to be presented in a rigorous form according to the manner of the ancients. (see Huygens 1659, 337)

Philosophers rightly mocked the concept of infinitesimals – Berkeley called them 'ghosts of departed quantities' – because of the loose and sometimes inconsistent way they were used by mathematicians. But mathematicians did not completely dispense with infinitesimals until forced to do so for mathematical reasons. And when they did it was part of a general revolution in mathematics that replaced geometry by arithmetic in the foundations of mathematics. We will see how this came about in the next two sections.

4.5 Complex Analysis

Calculus and imaginary numbers, though both had dubious origins, formed a powerful alliance in the eighteenth and nineteenth centuries. The most remarkable allies were the circular and exponential functions, which were studied separately in the seventeenth century and found to be expressible by infinite series. Around 1670 Newton discovered the series for the exponential function,

$$e^x = 1 + \frac{x}{1!} + \frac{x^2}{2!} + \frac{x^3}{3!} + \cdots,$$

and he and others rediscovered the sine and cosine series that had already been discovered in India:

$$\cos x = 1 - \frac{x^2}{2!} + \frac{x^4}{4!} - \frac{x^6}{6!} + \cdots,$$

$$\sin x = \frac{x}{1!} - \frac{x^3}{3!} + \frac{x^5}{5!} - \frac{x^7}{7!} \cdots.$$

Replacing x by ix in the series for e^x yields the miraculous formula

$$e^{ix} = \cos x + i \sin x,$$

discovered by Euler (1748). This formula not only allows sine and cosine to be expressed in terms of exponentials, namely

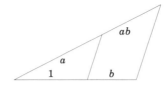

Figure 14 Product of lengths via similarity.

$$\cos x = \frac{1}{2}(e^{ix} + e^{-ix}), \ \sin x = \frac{1}{2i}(e^{ix} - e^{-ix}),$$

but also draws attention to their 'hyperbolic' analogues:

$$\cosh x = \frac{1}{2}(e^{x} + e^{-x}), \sinh x = \frac{1}{2}(e^{x} - e^{-x}).$$

This analogy may help to explain a wild conjecture of Lambert (1766, §82). Lambert introduced the hyperbolic functions and was no doubt aware of the formulas of spherical trigonometry, which involve sine and cosine. He may then have guessed that the analogous formulas involving hyperbolic sine and cosine describe trigonometry on a 'sphere of imaginary radius'. If so, this could explain his 1766 conjecture that non-Euclidean geometry may hold on an imaginary sphere. (See also Sections 6.2 and 6.5 for more about non-Euclidean geometry and Lambert's imaginary sphere.)

4.6 Philosophical Issues

Before discussing calculus, there is an important by-product of the Descartes (1637) book on algebraic geometry that should be mentioned – multiplication of lengths. As we mentioned in Section 1.3, there is a natural sum of lengths which is itself a length, but Euclid and his successors took the product of two lengths to be a box, which severely curtailed any algebra of lengths. But there is another way to define product of lengths, based on the proportionality of similar triangles (Figure 14).

To do so we choose a unit length, marked 1 in Figure 14, and construct the similar triangles containing the lengths *a* and *b* in the positions shown. Then it follows by proportionality that, just as the continuation of length 1 is length *b*, the continuation of length *a* is length *ab*. Descartes used this fact to define the product of lengths, thus giving lengths an algebraic structure like that of numbers. He also pointed out a simple geometric construction (also involving similar triangles) for the square root of any length. Thus Descartes (1637) swept away some of the ancient problems posed by the existence of irrational lengths: lengths did behave like numbers after all, and it was reasonable for the square root of any number to be a number.

But, as soon as one philosophical difficulty was removed, calculus created another. In fact, the invention of calculus disrupted the philosophy of mathematics perhaps more than any event since the discovery of irrational quantities. Philosophers rightly questioned the concept of infinitesimal, but mathematicians at first ignored their criticisms and continued to believe they could obtain calculus results by Euclid's methods (though they seldom actually did so). There was a stalemate which would not be broken until after 1800, when mathematicians had to concede that Euclid was, after all, not an adequate foundation for mathematics.

Intuition and logic. Intuition played a leading role in calculus (and still does), where 'infinitesimally close points on a curve' and 'infinitesimally thin strips' were used to set up equations for calculating tangents and areas. But, once the equations were found, the force of algebraic symbolism (in the algebra of infinitesimals) prevailed. Mathematicians believed in calculus because of its amazing success in solving problems in geometry and mechanics. They also thought they were on solid ground, believing that all their results could be obtained rigorously by the method of exhaustion.

Meaning and existence. Yet, as Berkeley pointed out, the existence of infinitesimals was highly dubious, so what explained their success? (Before Berkeley, Hobbes had made harsh criticisms of calculus and of the use of algebra in geometry. But he destroyed his credibility with mathematicians by proposing an untenable account of the circle – claiming that it contains only finitely many points – and claiming thereby to solve the ancient problem of 'squaring the circle'.)

Discrete and continuous. Then again, infinitesimals provided a new bridge (albeit rickety) between the discrete and continuous. The picture of the continuum was unclear, and perhaps contradictory, yet one could correctly calculate lengths and areas. So perhaps infinitesimals could help explain the nature of the continuum?

In the long run, the problems raised by infinitesimals were found to be problems about the nature of the continuum, and particularly its nature as a particular kind of infinite set. In the next two sections we will see how this problem, and its philosophical implications, unfolded.

5 Continuous Functions and Real Numbers

PREVIEW

Calculus could deal with many specific functions, but the general function concept remained vague. Around 1800, it became necessary to prove some general properties of continuous functions – surprisingly, in order to prove the fundamental theorem of algebra. After some dubious attempts to prove this theorem in the eighteenth century, Gauss proposed several proofs. The most convincing was one in Gauss (1816), which reduced the theorem to the special case of odd-degree polynomials. Such a polynomial $p(x)$ takes values that change 'continuously' from positive (for large positive x) to negative (for large negative x). It then seems obvious that the polynomial takes the value 0 somewhere, as required for the fundamental theorem.

Bolzano (1817) put his finger on the key assumption of Gauss's proof, the intermediate value theorem for continuous functions: if $f(x)$ is function that varies continuously from negative to positive as x varies, then $f(x) = 0$ for some value of x. Bolzano was able to give a satisfactory definition of continuity, but to prove the intermediate value theorem he had to assume a property of the real numbers, the *least upper bound property*: if S is a bounded set of real numbers, then S has a least upper bound.

It was not possible to prove the least upper bound property from Euclid's geometric concept of real number, which had been thought adequate until then. It was finally necessary to grapple with the unfinished business of reconciling the discrete with the continuous.

5.1 The Fundamental Theorem of Algebra

We saw in Section 3.1 that there is a formula for solving cubic equations, first published by Cardano in 1545. In the same book Cardano also published a formula for solving quartic (fourth-degree) equations. Both formulas express the solution in terms of the coefficients of the equation, the rational operations $+, -, \times, \div$, and square and cube roots. This raised the hope of similar solutions for higher-degree equations – solutions by radicals (n th roots), as they were called. This hope was eventually dashed in the 1820s by Abel and Galois, who showed that no such formulas exist for equations of fifth degree and higher.

But before solution by radicals was ruled out, Gauss had already suggested a different approach to polynomial equations: be content to prove *existence* of a solution rather finding it in a formula. Around 1800, Gauss gave several proofs in this style, the key ingredient of which was appeal to properties of continuous functions.

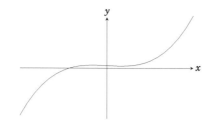

Figure 15 The graph of $y = x^3 - x + 2$.

The simplification made possible by an existence proof can be illustrated in the case of a cubic equation, say $x^3 - x + 2 = 0$. If one inspects the graph of $y = x^3 - x + 2$ (Figure 15) then it is immediately clear why $x^3 - x + 2 = 0$ for some value of x. The graph passes continuously from negative values (for large negative values of x) to positive values (for large positive values of x), and hence it somewhere takes the value 0.

The same argument applies to any polynomial of odd degree. In 1816 Gauss extended it to polynomials of any degree by some ingenious algebra that reduces the solution of an equation of degree $2n$ to the solution of a quadratic (where the solution may be complex) and an equation of degree n. By repeatedly halving the degree one reaches an equation of odd degree, and hence a solution exists. Thus Gauss's (1816) argument reduces the existence of solutions of any polynomial equation – the *fundamental theorem of algebra* as it is now called – to some algebra plus a seemingly obvious fact about continuous functions. Gauss was happy to assume facts about continuity, but in fact he stood on the brink of a new world of real analysis, in which the nature and properties of continuous functions were the focus of attention.

5.2 The Intermediate Value Theorem

Remarkably, Gauss's (1816) proof was immediately noticed, and its crucial idea about continuous functions formulated as a theorem by Bolzano (1817). This is what we now call the *intermediate value theorem*.

Intermediate Value Theorem. If f is a continuous function, defined for x with $a \leq x \leq b$, and if $f(a) < 0$ and $f(b) > 0$, then $f(c) = 0$ for some c between a and b.

To state the theorem, Bolzano had to define what 'continuous' means, and he came up with essentially the modern definition: f is *continuous at x* if, for each $\varepsilon > 0$ there is a $\delta > 0$ such that $|x - x'| < \delta$ implies $|f(x) - f(x')| < \varepsilon$. (Speaking less formally: we can make $f(x')$ as close as we please to $f(x)$ by choosing x' sufficiently close to x.) Then f is simply *continuous for x from a to b* if f is continuous at x whenever $a \leq x \leq b$.

Even when stated informally, it is not clear that this definition states what one wants 'continuous' to mean. One would prefer to say 'the graph of $y = f(x)$ is unbroken', or something like that – in fact, something like the intermediate value theorem. Bolzano's extraordinary insight led him to a definition that allows the vague global property of unbrokenness to follow from the precise local property of continuity at a point.

Bolzano certainly had the right definition of continuity to prove the intermediate value theorem, but his proof depended on an unproved property of numbers. This was the least upper bound principle, stating that any bounded set of numbers has a least upper bound. Bolzano could give only vague justification for this principle, because he had only a vague, geometric, conception of the number line. It was not possible to go further until the number concept had been defined in purely arithmetic fashion.

5.3 Definition of Real Numbers

Some decades after Bolzano, whose work received little attention, Dedekind experienced a similar dissatisfaction with intuitive arguments in calculus. In his 1872 booklet *Continuity and Irrational Numbers*, he wrote,

> As professor in the Polytechnic School in Zürich I found myself for the first time obliged to lecture on the elements of the differential calculus and felt more keenly than ever before the lack of a really scientific foundation for arithmetic. In discussing the notion of the approach of a variable magnitude to a fixed limiting value, and especially in proving the theorem that a magnitude which grows continually, but not beyond all limits, must certainly approach a limiting value, I had recourse to geometric arguments … a more careful investigation convinced me that this theorem, or any one equivalent to it, can be regarded in some way as a sufficient basis for infinitesimal analysis. It then only remained to discover its true origin in the elements of arithmetic and thus at the same time to secure a real definition of the essence of continuity. I succeeded Nov. 24, 1858.

By 'continuity', Dedekind means what we now call the *connectedness* of the real numbers; informally, that they have 'no gaps'. In this respect they contrast with the rational numbers, which have a gap at the position of $\sqrt{2}$ and many other places.

Now what, precisely, is a gap in the rational numbers? It is a partition of the rational numbers into two sets L and U, such that each member of U is greater than each member of L; L has no greatest member, and U has no least member. The gap at $\sqrt{2}$, for example, has upper set U consisting of all positive rationals with square greater than 2, and lower set L consisting of all the remaining rational numbers. Since $\sqrt{2}$ is not rational, L has no greatest member and U has no least. In general, each irrational corresponds to a gap in the rational numbers, which Dedekind called a *cut*.

This idea is clearly similar to the treatment of irrational quantities in Book V of Euclid's *Elements* (Section 2.4) except that infinite sets are no longer avoided. Dedekind's bold but simple idea was to use infinite sets as mathematical objects. A 'gap in the rationals' is then a meaningful mathematical object – a pair of sets L,U with the above properties – which we can take to define an irrational number. Thus the rationals and irrationals together form a number system without gaps, which we now call the real number system $\mathbb{R}$.

By basing the real numbers on rational numbers in this way, Dedekind had found their connectedness had its 'true origin in the elements of arithmetic'. Moreover, it is easy to show that the algebraic properties of the rational numbers (such as $a + b = b + a$ and $ab = ba$) carry over to the real numbers in a natural way. And by using the set concept, Dedekind found it equally easy to prove Bolzano's least upper bound principle, and hence provide an arithmetic foundation for real analysis.

To prove the least upper bound principle, we first represent each real number x by a set L_x of rational numbers which is bounded above and 'closed downward': that is, with the property that if $r \in L_x$ and $s < r$ then $s < L_x$. For an irrational number x, L_x is the L in the pair L,U that defines x; for a rational number x we take L_x to be the set of rationals $\leq x$. This representation has the convenient property that the ordering of real numbers corresponds to set containment: namely, $x \leq y$ if and only if $L_x \subseteq L_y$.

Then if we have a bounded set of real numbers x, the sets L_x are also bounded, and hence so is their union L. It follows that the real number l determined by the set L is the least upper bound of the numbers x.

5.4 Counter-intuitive Curves

By banishing geometric intuition from the foundations of analysis, Bolzano and Dedekind made it possible for *counter-intuitive* objects to be studied and accurately analysed. For some mathematicians, this was shocking development, and they 'turned away in horror and disgust from this awful plague' (to paraphrase a letter from Hermite to Stieltjes in 1893). But for others it was welcome, offering not only rigor but a way to actually sharpen one's intuition and see things that naive intuition could not.

The old formalism of infinitesimals was not sharp enough, because it suggests that a continuous function $y = f(x)$ is one for which an infinitesimal

Figure 16 The Koch polygon sequence.

change dx in x produces an infinitesimal change dy in y. The notation then leads us to believe that dy/dx exists; that is, the graph of a continuous $y = f(x)$ has a slope at each point. Indeed, Newton and other founders of calculus believed as much. But they were wrong: Bolzano and other nineteenth-century mathematicians discovered that there are continuous curves with no tangent at any point, and other counter-intuitive properties.

A lovely example of a curve without tangents, obtained as the limit of the sequence of polygonal curves, was given by von Koch (1904). The first five polygons of the sequence are shown in Figure 16.

It is, I think, 'intuitive' that this sequence has a continuous limit curve. But it can also be 'seen' that it has no tangents. It is clear from its construction that the limit curve can be divided into four pieces (corresponding to the four line segments in the second picture), each of which looks exactly the same as the whole curve when magnified by 3. But if the curve had a tangent at any point, the neighbourhood of that point would become straighter under magnification.

Many other examples of bizarre objects that intuition can be persuaded to grasp may be found in the book *In Search of Infinity* by Vilenkin (1995). They show that our geometric intuition is capable of much more than mathematicians thought, before the nineteenth century. Nevertheless, other challenges to intuition arose in the nineteenth century, as we will see in the next section.

5.5 Philosophical Issues

The proof of the intermediate value theorem is an infinite construction that captures the point c where $f(c) = 0$ by repeatedly halving the interval in which c ought to lie. It is not so different from a classical proof by exhaustion (assuming the existence of least upper bounds), except that it depends on being 'given' the function f. In Bolzano's time, it was not clear what this meant, which created the suspicion that the intermediate value theorem is purely an existence theorem, where the object c is claimed to exist but not constructed.

Actually, construction of c is not a problem when f is a polynomial, as in the fundamental theorem of algebra. But more existence theorems about continuous functions were to follow, and they became a bone of contention with 'constructivists' later in the nineteenth century. In any case, the more problematic part of Bolzano's proof is defining the real numbers so as to guarantee the least upper bound property and also their algebraic structure: in short, the *arithmetization of the line*. This led to an eruption of unforeseen philosophical problems, as we will see in the next two sections.

But, in the immediate wake of Bolzano and Dedekind, there were already several issues.

Intuition and logic. Contrary to the intuition that algebra is discrete, the fundamental theorem of algebra seems to involve continuity. And the intuition about continuous curves (and calculus in general) demands a deeper foundation, in an arithmetic theory of real numbers.

Meaning and existence. What does it mean to prove existence, without giving a formula for the object claimed to exist? What precisely are the real numbers, and what explains their completeness; that is, their closure under various infinite operations? (Dedekind's definition gives one answer; are there alternatives?)

Discrete and continuous. In particular, how best can we – avoiding the dubious means of infinitesimals – define the continuous (real numbers) in terms of the discrete (natural numbers)?

6 From Non-Euclidean Geometry to Arithmetic

PREVIEW

Meanwhile, Euclid had been found wanting in another respect. For hundreds, if not thousands, of years the parallel axiom had been considered an unnecessary blemish on Euclid's system. Many mathematicians had tried to prove it from Euclid's other axioms, but by 1800 hopes were fading. Some began to consider the previously unthinkable: a *non*-Euclidean geometry in which the parallel axiom was false.

This strange but wonderful geometry was at first explored only hypothetically. But Beltrami (1868) found models of it, showing that non-Euclidean geometry is just as consistent as Euclidean. With this discovery, Euclidean geometry lost its privileged position at the foundation of mathematics (and as the presumed geometry of physical space).

It was time to build a new foundation of mathematics, and arithmetic was ready to take the place of geometry. Dedekind had found an arithmetic definition of the real number line $\mathbb{R}$, on which it was possible to build a new foundation of geometry and analysis: the *real vector spaces* of Grassmann.

6.1 The Parallel Axiom

As we saw in the previous section, mathematicians in the first half of the nineteenth century lost faith in geometric intuition at the micro level, where it failed to grasp the local nature of the line. During roughly the same time period there was also a loss of faith in geometric intuition at the macro level, where it failed to grasp the global nature of lines – in particular, the behaviour of parallels.

Section 1.4 showed where the problem lies: in Euclid's parallel axiom, which implicitly claims the existence and uniqueness of parallels. The parallel axiom was a very useful axiom, used in the proof of many signature theorems of Euclid's geometry, such as the Pythagorean theorem. In fact, even the existence of squares depends on the parallel axiom, because unique parallels are needed to prove that the angle sum of a triangle is π, and hence that the angle sum of a quadrilateral is 2π. Another important consequence of the parallel axiom, which chimes well with our experience, is that figures of any size can have the same shape.

To many, theorems like these were more acceptable than the parallel axiom itself, and proofs that they imply the parallel axiom were found. The ultimate hope was that Euclid's other axioms – which are certainly more plausible than the parallel axiom – might be found to imply the parallel axiom, so that this 'blemish' on Euclid could be removed. Euclid's other axioms say uncontroversial things like: there is a unique line through any two points, lines are infinite, and if two triangles agree in two sides and the included angle then they agree in all sides and all angles.

If one takes the other axioms, and adds the axiom that parallels are not unique, the hope was that a contradiction would be discovered. The first to pursue the elusive contradiction was Saccheri (1733), in his book *Euclid*

Vindicated from Every Blemish. He found, not surprisingly, that these axioms imply that the angle sum of a triangle is less than π and the angle sum of a quadrilateral is less than 2π. He also found, more surprisingly, that two lines could be asymptotic; that is, they could approach each other arbitrarily closely without meeting. Saccheri found this 'abhorrent to the nature of straight lines', but still it was not a contradiction. In fact, it was an accurate glimpse of the non-Euclidean world.

6.2 Non-Euclidean Geometry

In the first decades of the nineteenth century, mathematicians began to explore the hypothetical non-Euclidean world less sceptically. Gauss himself was aware of the implications of non-unique parallels, but was afraid to publish them. It was left to Bolyai and Lobachevsky, in the 1820s, to independently develop and publish the basic theorems. There was also a hint, in the work of Minding in the 1830s, that the formulas of non-Euclidean trigonometry made sense on certain saddle-shaped surfaces. These formulas, which are like those of spherical trigonometry except that sine and cosine are replaced by their hyperbolic analogues, hold if 'lines' were taken to be geodesics – curves of shortest length – on the surface. But Minding's discovery fell short of being a full realization of non-Euclidean geometry, because his surfaces were incomplete – they did not admit infinite 'lines' in every direction.

The difficulties of modelling non-Euclidean geometry were finally overcome by Beltrami (1868) by relaxing the definition of 'distance'. In fact, Beltrami found several realizations, or models, of non-Euclidean geometry in which 'lines' were quite elegant and natural. Perhaps the easiest to grasp is the *conformal disc model* shown in Figure 17.

In this figure, the 'plane' is the interior of the disc, so its 'points' are points inside the disc boundary. Its 'lines' are circular arcs perpendicular to the disc boundary and 'angles' are the actual angles between 'lines'. (This is why the model is called conformal, which means that it faithfully represents angles.) The concept of distance is via a certain formula I will not state, but one can get a general impression of 'distance' as follows.

The figure shows many 'triangles', each of which has the same shape because it has angles $\pi/2, \pi/3, \pi/7$ (thus, like all non-Euclidean triangles, they have angle sum $< \pi$). Since figures of the same shape have the same size in non-Euclidean geometry, these triangles all have the same non-Euclidean size. In particular, one sees that the non-Euclidean plane and its 'lines' are infinite, because each line passes through infinitely many triangles on its way to the disc boundary. One can also guess that the 'lines' are curves of shortest 'length',

Figure 17 The conformal disc model of the non-Euclidean plane.

because it appears that a 'line' gives the shortest path between any two points, if measured by the number of triangles it passes through.

6.3 The Impact of Non-Euclidean Geometry

Non-Euclidean geometry finally killed any chance of deducing the parallel axiom from Euclid's other axioms, because the other axioms hold in Beltrami's model but the parallel axiom does not. One can see the failure of the parallel axiom directly in Figure 18, where *l* is a 'line', *P* is a point outside *l*, and two 'lines' *m* and *n* pass through *P* without meeting *l*.

Of course, there are models in which the parallel axiom does hold, so we see that the axiom is independent of the other axioms, and it can be replaced by the *non-Euclidean parallel axiom*, stating the existence of multiple parallels, without fear of contradiction. The immediate effect of this discovery – the first independence proof in mathematics – was increased distrust of Euclid's geometry as a foundation of mathematics, and increased support for arithmetic. In fact, as we will see in the next section, an arithmetic approach to geometry based on real vector spaces was ready to be rolled out, though few mathematicians were ready to understand it.

Certainly, arithmetization of analysis was gaining support. Thanks to the influence of Weierstrass, who had proposed an arithmetic definition of real numbers independently of Dedekind in 1864, arithmetical proofs of the basic

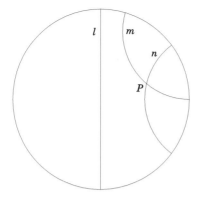

Figure 18 Failure of the parallel axiom.

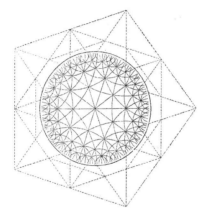

Figure 19 The Schwarz tessellation.

theorems on continuous functions (and hence of the fundamental theorem of algebra) began to circulate in the 1870s.

Another remarkable discovery, by Poincaré and Klein in the 1880s, was that non-Euclidean geometry was already present in mathematics. They noticed that there was non-Euclidean periodicity in the behaviour of functions on the unit disc of complex numbers, and that the phenomenon had been observed earlier, by Gauss, Riemann, and Schwarz, without realizing its geometric significance. Schwarz (1872) even produced a diagram (Figure 19) of the disc that is none other than a tessellation of the conformal disc model by congruent non-Euclidean triangles!

These discoveries cemented the position of non-Euclidean geometry in mathematics, and indeed in the arithmetic of complex numbers.

Outside geometry, the independence of the parallel axiom foreshadowed other independence proofs in mathematics. In cases where a sentence σ is not known to be a consequence of axioms $\sum$, one might hope to prove σ

independent of $\sum$ by finding two models of $\sum$: one satisfying σ and the other satisfying the negation of σ. We will mention examples later.

The independence of the parallel axiom was also a harbinger of *reverse mathematics*, a branch in which one seeks the 'right axiom' to prove a given theorem. The parallel axiom is the right axiom to prove, say, the Pythagorean theorem because the two are equivalent – given Euclid's other axioms – but neither can be proved from these axioms alone. Indeed, the parallel axiom is the right axiom to prove the host of theorems shown to be equivalent to the parallel axiom before its independence was known.

As we will see in Section 9, the phenomenon of independence became a big story in twentieth-century mathematics, changing the way we think about truth and proof.

6.4 Arithmetization of Geometry

In an obscure publication called *Die Ausdehnungslehre* (extension theory), Grassmann (1844) introduced a revolutionary approach to geometry, based on what we now call real vector spaces. Unfortunately, Grassmann's style was impenetrable to his contemporaries, even when he published a simplified version in 1847, emphasizing the role of the inner product. The first to appreciate his work was Peano (1888), who gave an axiom system for real vector spaces, one of the first modern axiom systems. Interestingly, Grassmann credited the germ of the idea to a sketchy manuscript of Leibniz from 1679 called *Characteristica Geometrica*. If Grassmann really saw vector spaces in the *Characteristica* he was the only one to do so (others have thought that the *Characteristica* foreshadows the very different subject of combinatorial topology). At the very least, Grassmann deserves most of the credit for recognizing the vector space concept, and for seeing what must be added to make it truly geometric.

Today we see vector spaces all over mathematics, so most mathematicians are familiar with the basic ideas. There is a set V of objects called vectors, which can be multiplied by real numbers and added. They include a zero vector $\mathbf{0}$ and, for each vector $\mathbf{a}$ a vector $-\mathbf{a}$ called the negative of $\mathbf{a}$. Vectors are governed by the following axioms. The first group say that the vector sum $+$ behaves like ordinary sum; that is, for any $\mathbf{u}, \mathbf{v}, \mathbf{w}$ in V:

$$\mathbf{u} + \mathbf{v} = \mathbf{v} + \mathbf{u},$$
$$\mathbf{u} + (\mathbf{v} + \mathbf{w}) = (\mathbf{u} + \mathbf{v}) + \mathbf{w},$$
$$\mathbf{u} + \mathbf{0} = \mathbf{u},$$
$$\mathbf{u} + (-\mathbf{u}) = \mathbf{0}.$$

And the second group say that multiples of vectors behave like multiples of numbers; that is, for any a,b in $\mathbb{R}$ and $\boldsymbol{u}, \boldsymbol{v}$ in V,

$$
\begin{aligned}
a(b\boldsymbol{u}) &= (ab)\boldsymbol{u}, \\
1\boldsymbol{u} &= \boldsymbol{u}, \\
a(\boldsymbol{u} + \boldsymbol{v}) &= a\boldsymbol{u} + a\boldsymbol{v}, \\
(a + b)\boldsymbol{u} &= a\boldsymbol{u} + b\boldsymbol{u}.
\end{aligned}
$$

It is clear from these axioms that $\mathbb{R}$ itself is a real vector space, but a more interesting example is $V = \mathbb{R}^2$, with sums and multiples of the vectors (x,y) defined, for each $x,y,a \in \mathbb{R}$, by

$$
\begin{aligned}
(x_1,y_1) + (x_2,y_2) &= (x_1 + x_2, y_1 + y_2) \\
a(x,y) &= (ax, ay).
\end{aligned}
$$

$\mathbb{R}^2$ has considerable geometric content. One can define lines, parallel lines, and a relative concept of length along a given line. For example, one can say that the multiples $t\boldsymbol{v}$ of a non-zero vector $\boldsymbol{v}$ form a line through $\boldsymbol{0}$, and that $2\boldsymbol{v}$ is twice as far from $\boldsymbol{0}$ as $\boldsymbol{v}$. However, there is no concept of distance between arbitrary points. To obtain the natural concept of distance, Grassmann introduced the *inner product*:

$$
(x_1,y_1) \cdot (x_2,y_2) = x_1 x_2 + y_1 y_2.
$$

From this definition it follows that if $v = (x,y)$, then

$$
\boldsymbol{v} \cdot \boldsymbol{v} = x^2 + y^2 = |v|^2,
$$

where $|v|$ is the distance of $|v|$ from the origin $\boldsymbol{0}$ given by the Pythagorean theorem. As Grassmann (1847) pointed out, his inner product is essentially equivalent to the Pythagorean theorem. The concept of angle is also inherent in the inner product, because

$$
\boldsymbol{u} \cdot \boldsymbol{v} = |\boldsymbol{u}||\mathbf{v}|\cos\theta,
$$

where θ is the angle between the lines from $\boldsymbol{0}$ to $\boldsymbol{u}$ and $\boldsymbol{v}$.

6.5 Vector Geometry

The previous section shows that the plane of Descartes (Section 4.2), in which points are ordered pairs (x,y) of reals x,y and distance is given by the Pythagorean theorem, has the same ingredients as the vector space $\mathbb{R}^2$ with Grassmann's inner product. The difference is that the algebra of Descartes, which involves arbitrary polynomials in x and y, goes much further than Euclid. Grassmann's algebra of the vector space $\mathbb{R}^2$, involving only sums and real multiples of vectors, and the inner product, is at just the right level to capture Euclid's geometry. For this reason, the vector space $\mathbb{R}^2$ with Grassmann's inner product is called the *Euclidean plane*.

Grassmann's vector geometry also generalizes to any number of dimensions with no extra effort. By working with n-tuples $(x_1, x_2, \ldots, x_n)$ of real numbers, one can do geometry in any number of dimensions without the need for visualization, thus breaking the dimension barrier that held back the Greeks.

In another direction, it is useful to generalize the definition of 'distance' by generalizing the concept of inner product. A famous example is *Minkowski space* defined by Minkowski (1908) as a setting for Einstein's special theory of relativity. This is the space $\mathbb{R}^4$ of 4-tuples (t, x, y, z) with distance derived from the inner product

$$(t_1, x_1, y_1, z_1) \cdot (t_2, x_2, y_2, z_2) = -t_1 t_2 + x_1 x_2 + y_1 y_2 + z_1 z_2.$$

Thus the Minkowski distance of (t, x, y, z) from the origin has square equal to $-t^2 + x^2 + y^2 + z^2$, which is sometimes negative. Minkowski space is the natural space for physics, where t stands for time and x, y, z are variables for the three dimensions of space. But mathematically it is equally interesting to look at the 3-dimensional space obtained by dropping the z coordinate.

In this space there is a 'sphere of radius $\sqrt{-1}$', consisting of the points $\mathbf{v} = (t, x, y)$ with $|\mathbf{v}|^2 = -1$, that is

$$-t^2 + x^2 + y^2 = -1, \text{ or } t^2 - x^2 - y^2 = 1.$$

This is none other than the hyperboloid, shown in Figure 20. In terms of Minkowski distance, the geometry on this hyperboloid is none other the non-Euclidean geometry of Beltrami! In Figure 20 (based on one due to Konrad Polthier of the Free University of Berlin) we have indicated how triangles in conformal disc model (Figure 17) correspond to triangles on the hyperboloid that are equal in the sense of Minkowski distance.

Minkowski space gives substance to the wild idea of Lambert from back in 1766 (mentioned in Section 4.5), that non-Euclidean geometry might hold on a sphere of imaginary radius.

Euclidean and Minkowski spaces show that $\mathbb{R}$ is a natural and convenient foundation for geometry, both Euclidean and non-Euclidean. The next question is: what is a foundation for $\mathbb{R}$? In Section 8 we will see two answers to this question. But first we need to take a closer look at $\mathbb{R}$ itself, in Section 7, to appreciate how subtle its foundation may be.

Figure 20 The hyperboloid model of non-Euclidean geometry.

6.6 Philosophical Issues

A key step in Beltrami's modelling of non-Euclidean geometry was his decision to generalize the concept of length. He was emboldened to do this by his reading of Riemann (1854), a ground-breaking work in the foundations of geometry that was first published in 1868. In a sweeping new approach to geometry (now known as Riemannian geometry), Riemann generalized the arithmetized geometry of Descartes from the plane to curved spaces of any dimension n. The points of the space, instead of being represented by ordered pairs of numbers, were represented by ordered n-tuples $(x_1, x_2, \ldots, x_n)$. And distance, instead of being given by the 'Pythagorean' formula, was obtained by calculus from an 'infinitesimal Pythagorean' formula. The 'infinitesimal Pythagorean' property means that the geometry of the space (as manifested by the angle sum of a triangle, for example) approaches Euclidean geometry in small regions, but may diverge from it in large regions. An example is the geometry of the sphere, where small triangles have angle sum close to π, but large triangles have angle sum much greater than π.

The sphere is an example of a space with constant positive curvature. Riemann looked briefly at spaces of constant negative curvature, but it was Beltrami who noticed that such spaces exhibit non-Euclidean geometry, and that their 'lines' (curves of shortest length) can be modelled simply by circular arcs, as in Figure 17. Thus Riemannian geometry is a thoroughgoing arithmetization of geometry, pointing the way to an arithmetic basis for Euclidean, non-Euclidean, and all kinds of curved geometries.

The revolution wrought by Riemann and Beltrami raised several issues about the nature of geometry and its place in mathematics.

Intuition and logic. Models show that Euclidean and non-Euclidean geometries are equally sound, and the real numbers provide a foundation for both (and also for calculus and mathematical physics). It remains to find a good foundation for the real numbers; hopefully based on the arithmetic of natural numbers. As we know, Dedekind found one way to do this (Section 5.3).

Meaning and existence. But the real numbers involve more than arithmetic; namely, some assumption about infinity. How much is it necessary, and legitimate, to assume?

Discrete and continuous. Bridging the gap between discrete and continuous is essentially the problem of defining real numbers in terms of natural numbers. Hence the possibility of bridging the gap (and arithmetizing geometry) depends on what it is legitimate to assume about infinity.

7 Set Theory and Its Paradoxes

PREVIEW

Dedekind's definition of real numbers opened a new era in mathematical thought, in which infinite sets were viewed as mathematical objects. This was unwelcome to many mathematicians, who took the ancient view that actual infinity was unacceptable. Indeed, in some quarters, resistance to actual infinity continues to this day.

Even more unwelcome was the discovery of Cantor (1874) that the real numbers form an *uncountable* set – one that cannot be finessed as a merely potential infinity rather than actual.

But uncountability was not the last straw. Cantor (1891) found a simple generalization of his argument that shows there is no largest set. Indeed, for any set S, the subsets of S are more numerous than members of S.

It follows that there is no 'set of all sets' and therefore not every property is realized by a set. In particular, there is no set that realizes the property 'X is a set'. Even Dedekind was worried by this development.

7.1 Before Cantor

In Section 2.3 we mentioned that the Greeks were suspicious of infinity, and explained how they avoided it as far as possible. Nevertheless, the Greeks certainly used infinite processes, and very ingeniously at that. Their avoidance in practice meant considering arbitrary finite parts of an infinite totality or process – the 'potential' rather than the 'actual' infinite.

The actual infinite was avoided because of its seemingly paradoxical properties, such as apparent violation of the principle that 'the whole is greater than the part'. This paradox resurfaced with the revival of Greek learning in Medieval and Renaissance times. For example, Galileo pointed out the correspondence

1	2	3	4	...
$\updownarrow$	$\updownarrow$	$\updownarrow$	$\updownarrow$	
1^2	2^2	3^2	4^2	

which seemingly shows the whole collection of positive integers to be no greater than its part consisting of squares.

Until the late nineteenth century this kind of relationship – a one-to-one correspondence between a part and the whole of some collection – was considered paradoxical. But eventually one-to-one correspondence came to be seen as the right way to compare infinite sets. Indeed, Dedekind (1888) took the defining property of an infinite set to be that its whole can be put in one-to-one correspondence with a part, thus turning a paradox into a definition. Two sets that can be put in one-to-one correspondence with one another are called *equinumerous* or of the same cardinality.

Many sets are equinumerous with the set $\mathbb{N}$ of positive integers. For example:

1 The set $\mathbb{Z}$ of all integers, equinumerous with $\mathbb{N}$ via the correspondence

1	2	3	4	5	6	7	...
$\updownarrow$	$\updownarrow$	$\updownarrow$	$\updownarrow$	$\updownarrow$	$\updownarrow$	$\updownarrow$	
0	1	-1	2	-2	3	-3	

2 The $\mathbb{Q}^+$ of positive rationals, via the correspondence

1	2	3	4	5	6	7	8	9	...
$\updownarrow$	$\updownarrow$	$\updownarrow$	$\updownarrow$	$\updownarrow$	$\updownarrow$	$\updownarrow$	$\updownarrow$	$\updownarrow$	
$1/1$	$2/1$	$1/2$	$3/1$	$1/3$	$4/1$	$3/2$	$2/3$	$1/4$	

where the bottom line lists the distinct fractions m/n in groups: first those with $m + n = 2$, then those with $m + n = 3$, those with $m + n = 4$, and so on.

3 The set $\mathbb{Q}$ of all rationals can then be shown equinumerous with $\mathbb{N}$ by the same trick used to list $\mathbb{Z}$: list 0 first, then alternately list the positive and negative form of each positive rational.

These results show that $\mathbb{Z}$, $\mathbb{Q}^+$, and $\mathbb{Q}$ can all be viewed as 'potential' infinities like $\mathbb{N}$. An even stronger result along the same lines was found by Dedekind in 1874: the set of all algebraic numbers (solutions of polynomial equations with integer coefficients) is equinumerous with $\mathbb{N}$. So it too can be viewed as a 'potential' infinity. With these results most of the ancient fears about actual infinity could be dismissed – because every infinity seemed to be merely 'potential' – but there was a big surprise just around the corner.

7.2 Cantor's Diagonal Argument

Cantor (1874) proved that the real numbers are not equinumerous with the positive integers. That is, any pairing of positive integers with real numbers,

1	2	3	4	5	6	7	$\ldots$,
$\updownarrow$	$\updownarrow$	$\updownarrow$	$\updownarrow$	$\updownarrow$	$\updownarrow$	$\updownarrow$	
x_1	x_2	x_3	x_4	x_5	x_6	x_7	

fails to include all the real numbers. In fact, given any list $x_1, x_2, x_3, x_4, \ldots$ of real numbers we can explicitly describe a real number x not on the list. The 1874 proof was not easy to follow – especially for a mathematical community completely unprepared for it – but Cantor (1891) gave another proof which obtains the 'witness' x with maximum clarity. This is the famous (or, to some, notorious) diagonal argument.

There are many ways in which the real numbers $x_1, x_2, x_3, x_4, \ldots$ can be given, but to be specific we will suppose they are given as infinite decimals. We will also ignore digits before the decimal point, so we can imagine the numbers displayed as in Figure 21, showing just the digits after the decimal point:

x_1	**1**	1	1	1	$\ldots$
x_2	0	**2**	0	1	$\ldots$
x_3	7	7	**7**	7	$\ldots$
x_4	0	0	0	**0**	$\ldots$
$\vdots$					
x	**2**	**1**	**1**	**1**	$\ldots$

Figure 21 The diagonal construction.

Figure 21 also shows the first few digits of x. We ensure that $x \neq$ each x_n by being different in the n th decimal place (and not using the digits 0 and 9 in x, because numbers with these digits can be the same even though their digits are different – for example $0 \cdot 4999 \ldots = 0 \cdot 5000 \ldots$). Specifically, we define x by

$$n\text{th digit of } x = \begin{cases} 2 & \text{if } n\text{th digit of } x_n \text{ is } 1 \\ 1 & \text{if } n\text{th digit of } x_n \text{ is not } 1. \end{cases}$$

Thus x is different from all the numbers $x_1, x_2, x_3, x_4, \ldots$, hence the given list does not include all real numbers. Because of this, we say that the set of all real numbers is uncountable – a countable set being one whose members can be paired with the positive integers.

Since only countable sets can be considered 'potentially' infinite, the set $\mathbb{R}$ of real numbers is unavoidably an actual infinity.

Cantor's argument is called 'diagonal' because it involves just the digits on the diagonal of the table, shown in bold in the figure. Thus we need only inspect a finite amount of each decimal expansion – namely, the first n digits of x_n – to calculate the n th digit of x. I mention this to dispel the common misconception that the diagonal argument merely proves the *existence* of a number x not on the given list $x_1, x_2, x_3, x_4, \ldots$. In fact it shows that x is just as constructible as the numbers $x_1, x_2, x_3, x_4, \ldots$ themselves.

7.3 Higher Infinities

The essence of the diagonal argument is to say: given a real number x_n paired with each natural number n, we can define a real number $x \neq$ each x_n by the rule

$$n\text{th digit of } x = \begin{cases} 2 & \text{if } n\text{th digit of } x_n \text{ is } 1 \\ 1 & \text{if } n\text{th digit of } x_n \text{ is not } 1, \end{cases}$$

because this makes $x \neq x_n$ in the n th decimal place. A similar argument shows that there can be no list $S_1, S_2, S_3, S_4, \ldots$ of all subsets of $\mathbb{N}$. Because for any such list we can define a set $S \neq$ each S_n by the rule

$$n \in S \Leftrightarrow n \notin S_n,$$

since this rule makes S different from S_n with respect to the number n.

In his 1891 paper Cantor took this train of thought to the end of the line, showing that every set X has more subsets than members. To see why, suppose that each member x of X is paired with a subset S_x of S. But then we can define a subset S of X different from each S_x with respect to the member x. Namely, let

$$x \in S \iff x \notin S_x.$$

Thus a collection of subsets S_x paired with members x of X does not include all subsets of X.

What is slightly alarming about this result is that it shows that there is no set of all sets. If there were such a set, X say, then the collection of its subsets would be a set bigger than the set of all sets, which is clearly contradictory. Cantor soon noticed this consequence of the diagonal argument, but he remained calm. (Dedekind, however, was surprised by this development, and he delayed publication of a second edition of his book Dedekind [1888] as a result.) History would show that it is natural for the collection of all sets not to be a set, much as it is natural for the collection of all positive integers not to be a positive integer. However, at the time there was alarm because of a previous belief that every property should be realized by a set. After Cantor's discovery, it was clear that there are exceptions to this belief, such as the property of being a set.

7.4 Aftermath of the Diagonal Argument

The diagonal argument was a pivotal discovery in the foundations of mathematics. In one direction it led, seemingly inexorably, to higher realms of infinity – and towards new kinds of paradox. In another direction it led to the discovery of limitations in formal mathematics: incompleteness and algorithmic unsolvability, as we will see in the next two sections. And it also provoked reaction from a new breed of sceptics, whose scepticism was directed not just at infinity but at logic itself.

Cantor's discovery that there is no largest set, and that not every property is realized by a set, led to the development of set theory, a systematic study of infinite sets in general and $\mathbb{R}$ in particular. The uncountability of $\mathbb{R}$ was merely the first of many confounding discoveries about $\mathbb{R}$, showing that the number line we thought we knew is more complicated than anyone imagined. In fact, to answer many questions about $\mathbb{R}$ we have to make assumptions about the whole universe of sets.

On the other hand, the simple and computational nature of the diagonal argument makes it applicable to the most down-to-earth 'potential' kinds of infinity. For example, it shows that there is no computable list of all computable numbers. Because if $x_1, x_2, x_3, \ldots$ is a computable list of computable numbers then the diagonal number x is also computable – and it is not on the list. This result has profound implications for what can be computed, in all areas of mathematics.

Not surprisingly, the diagonal argument and its implications were not welcome to all mathematicians. Those already suspicious of infinity, such as Cantor's Berlin colleague Kronecker, became even more vehement in their opposition to sets such as $\mathbb{R}$. Weierstrass was more sympathetic, but he persuaded Cantor to tone down the uncountability aspect of the proof and to emphasize instead a positive outcome: a new and elementary proof of the existence of non-algebraic numbers – the so-called *transcendental* numbers. (This follows immediately from the uncountability of $\mathbb{R}$ and Dedekind's result that the set of algebraic numbers is countable.)

In the years that followed, Cantor's ideas gradually gained the support of the majority, thanks to the support of eminent mathematicians such as Hilbert. Some opposition remained, but it became more nuanced as it was gradually understood how much mathematics has to be sacrificed if various kinds of infinity are rejected. Among the extreme rejectionists, the most prominent are the constructivists, who accept existence proofs only when they provide a construction of the object claimed to exist. This attitude has had a positive influence even on mathematicians who do not share it. For example, we now know how to construct many objects, such as solutions of polynomial equations, first shown to exist by non-constructive arguments.

7.5 Philosophical Issues

It may be worth mentioning that Cantor's first proof of the uncountability of $\mathbb{R}$ was also a diagonal argument, but in a less transparent form. It had the same essence: constructing a real number x step by step and making x different from the real number x_n at step n. However, the construction was specific to real numbers, so the spectacular generalization to arbitrary sets found by Cantor (1891) was new. There is another uncountability proof, specific to $\mathbb{R}$, which is more revealing in my opinion. It goes like this.

Given real numbers $x_1, x_2, x_3, \ldots$, we cover each x_n by an interval of length 2^{-n}. Then the total length of the number line covered is at most

$$\frac{1}{2} + \frac{1}{4} + \frac{1}{8} + \ldots = 1,$$

which is surely not the whole line. So the points $x_1, x_2, x_3, \ldots$ cannot include all real numbers. This proof can be refined to give a specific number x not covered. We choose the first (binary) digit of x to avoid the first interval, then the second digit to avoid the second interval, and so on – at which stage it becomes clear that this construction is another diagonal argument.

There is in fact another route to uncountability, through Cantor's theory of ordinal numbers. Interesting though this is, it raises another set of issues that we do not have space to discuss here. The existence of uncountable sets raises enough issues on its own.

Intuition and logic. How much understanding of infinity is humanly possible? Dedekind (1888, §66) made an audacious attempt to prove that infinite sets exist (following a similar attempt by Bolzano 1851, §13). It is based on the fact that only an infinite set can be equinumerous with a proper subset of itself:

> *Theorem.* There exist infinite systems.
> *Proof.* My own realm of thoughts, i.e. the totality S of all things which can be objects of my thought, is infinite. For if s signifies an element of S, then is the thought s', that s can be an object of my thought, itself an element of S. If we regard this as a transform $\phi(s)$ of the elements s then has the transformation ϕ of S, thus determined, the property that the transform S' is part of S; and S' is certainly a proper part of S, because there are elements in S … which are not contained in S'.

Spoilsports will no doubt object that in his lifetime Dedekind had only a finite number of thoughts, so there must be something wrong with this 'proof'. But who knows what the realm of Dedekind's thoughts really is? Dedekind's argument had some eminent supporters, such as Russell (1903, 357).

Bolzano and Dedekind considered only countable sets, but we might also shoot for uncountable sets. Is our apparent intuition of continuity an intuition of uncountability and/or actual infinity?

Meaning and existence. Before Cantor, the question of infinity was simple: do we accept actual infinity or not? After Cantor, the question became more complex: how much infinity do we accept? When infinity is found to have infinitely many possible levels, many different levels of acceptance are possible. Some mathematicians accepted only countably infinite sets (the potential infinite), others accepted $\mathbb{R}$ but not all subsets of $\mathbb{R}$, and so on. The French mathematicians Borel, Baire, Lebesgue, and Hadamard had a lively debate about this in 1905, which may be read in Ewald (1996, II:1077–86). In response to Borel's view that certain procedures, such as making uncountably many choices, were 'outside mathematics', Hadamard retorted:

> From the invention of the infinitesimal calculus to the present, it seems to me, the essential progress in mathematics has resulted from successively annex-ing notions which, for the Greeks or the Renaissance geometers or the predecessors of Riemann, were 'outside mathematics' because it was impos-sible to describe them. (see Ewald 1996, 1084)

Discrete and continuous. The uncountability of $\mathbb{R}$ conclusively shows that mathematics involving $\mathbb{R}$ – such as analysis – touches on set theory. At the very least, the project of arithmetizing geometry and analysis must go some distance beyond pure arithmetic: it must include some assumptions about infinite sets.

In the next section we will look at the standard axiom system of arithmetic, and catch a glimpse of the axioms about sets that lie beyond.

8 Formal Systems

PREVIEW

The sudden appearance of set theory and its paradoxes was not the only philosophically significant development in late nineteenth-century mathematics. Another was the emergence of formal logic and, more generally, formal systems for mathematics. Beginning with appearance of Boole's *Mathematical Analysis of Logic* in 1847, logic and mathematics were translated into symbolism in which deductions were essentially calculations.

A century ago, systems ready for formalization were known for arithmetic (Peano 1889), geometry (Hilbert 1899), and set theory (Zermelo 1908). There was also a formal system for logic, due to Frege (1879).

Formal systems revived an old dream of Leibniz: a *calculus ratiocinator* by which the truth of any proposition could be decided by calculation. It is true that the theorems of a formal system $\sum$ are obtainable by calculation, so if τ is a theorem of $\sum$ we will eventually observe this fact by systematically generating theorems. However, it is not clear whether

1 all truths expressible in $\sum$ are theorems, (Completeness)
2 some rule decides, for each τ, whether τ is a theorem, (Decidability)
3 if $\sum$ proves τ then $\sum$ does not prove the negation of τ. (Consistency)

Nevertheless, these are all questions about the outcomes of finite computations, like questions of number theory. So one might hope to settle them by elementary means (which Hilbert called 'finitary'). In particular, by a programme for settling the consistency question for formal set theory, Hilbert hoped to remove all doubt about the use of infinity in mathematics.

8.1 Hilbert

In the 1890s Hilbert reinvigorated Euclid's geometry in a thorough study of geometric axioms and their relation to algebra and the real numbers. Building on work of some of his predecessors, such as von Staudt and Pasch, he

1 filled the gaps in Euclid's system, by explicitly stating axioms that Euclid had used unconsciously,
2 grouped the axioms into conceptually different types: incidence, order, congruence, circle intersection,
3 derived the algebraic properties of sum and product from incidence axioms, and their order properties from the order axioms,
4 added two axioms not needed for geometry but needed to derive the properties of the real number line: the *Archimedean axiom* (stating that there are no infinitesimals), and a *completeness axiom* (stating that there are no gaps, in the sense of Dedekind).

It is evident from the inclusion of Archimedean and completeness axioms that Hilbert was interested in the real numbers as much as he was interested in geometry. In fact, in his 1899 book *Grundlagen der Geometrie* (Foundations of geometry) he also included an ingenious new construction of $\mathbb{R}$ from the axioms of non-Euclidean geometry (obtained by replacing the parallel axiom).

The derivation of $\mathbb{R}$ from geometric axioms has not been used much in mathematics, but it has made a curious contribution to philosophy. Hartry Field (1980), in his *Science without Numbers*, adapted Hilbert's result to derive the structure of the real numbers from axioms about physical space. He used this derivation to claim that $\mathbb{R}$ is dispensable in science, and hence that no one need assume its existence. But, to most mathematicians, assuming that the *structure* of $\mathbb{R}$ exists (in physical space or in any abstract realm) is the same as assuming the existence of $\mathbb{R}$ itself. And assuming that physical space is archimedean, complete, and Euclidean – merely to avoid assuming the existence of $\mathbb{R}$ – seems to defy physics as much as it defies mathematics.

8.2 The Systems of Peano and Zermelo

The more usual way to arrive at $\mathbb{R}$ in mathematics is through axioms for the natural numbers and sets. Axioms for the natural numbers were given by Dedekind (1888) and Peano (1889), in a system now called *Peano arithmetic*, or PA. Peano acknowledged an idea of Grassmann (1861) that the crucial concept of arithmetic is induction. 'Induction' here is not the rough idea of drawing a general conclusion from particular cases, but an ironclad guarantee that the general conclusion is correct. This kind of induction is often called complete induction because, for natural numbers, it covers all cases.

What makes it possible to prove a general claim $\sigma(n)$ about an arbitrary natural number n is that n can be reached from 0 by repeatedly adding 1. Therefore, if we can prove that $\sigma(0)$ holds, and that $\sigma(m + 1)$ holds whenever

$\sigma(m)$ holds, then we can be sure that $\sigma(n)$ holds for each natural number n. Peano built this idea into his axiom system in two ways.

- After including names 0 and S for the initial number and the successor function $S(m) = m + 1$, with appropriate properties, he gave inductive definitions of sum and product:

$$l + 0 = l, \, l + S(m) = S(l + m),$$

$$l \cdot 0 = 0, \, l \cdot S(m) = l \cdot m + l.$$

It follows from these definitions that the functions $+$ and $\cdot$ are defined for all natural numbers. For example, the first equation in the definition of $+$ defines $l + m$ for $m = 0$; the second defines $l + S(m)$ once $l + m$ is defined, and hence defines $l + n$ for all natural numbers n, by complete induction. Likewise, the second pair of equations defines $l \cdot n$ for all n, given that $+$ is already defined.

- The definitions of $+$ and $\cdot$ enable all particular facts about sum and product for the numerals $0, S(0), SS(0), \ldots$ to be derived, by substituting in the defining equations. But to prove general facts, such as $l + m = m + l$, Peano provides the induction axiom for each property σ: If $\sigma(0)$ and if $\sigma(m) \Rightarrow \sigma(S(m))$ for all m, then $\sigma(n)$ holds for each n.

(An equivalent induction axiom is that every set of natural numbers has a least member, or that a descending sequence of natural numbers is finite. In the latter form induction goes back to Euclid.)

Zermelo gave axioms for set theory in 1908. We omit the details, but they are similar to the Dedekind or Peano axioms in spirit, as Zermelo acknowledged. They assert the existence of a starting set $\emptyset$ (the empty set, which can be viewed as 0), operations for building further sets (which, among other things, allow successors of 0 to be built), and an axiom of infinity stating the existence of a set including 0 and all its successors. There is also an axiom of foundation that is similar to induction. Indeed, if the axiom of infinity is omitted, Zermelo's set theory has essentially the same content as the Peano axioms. So set theory in a sense is 'number theory + infinity'.

Set theory is an extremely powerful system, capable of covering virtually all of mainstream mathematics. This is because it has set construction principles – such as forming the set $\mathcal{P}(X)$ of all subsets of a set X – that cause explosive growth once an infinite set X is present. Beginning with Hilbert in the 1930s, there has been interest in systems with milder set construction principles, tailored to analysis. In these systems it turns out that we can measure the 'strength' of various theorems of analysis by the set construction principles needed to prove them. (It happens surprisingly often that we can find the 'right set construction axioms' to prove theorems of analysis, rather like finding the parallel axiom to be the 'right axiom' to prove many theorems of geometry. This phenomenon is studied in the new field of reverse mathematics mentioned in Section 6.3.)

8.3 Frege's System for Logic

So far we have been vague about how axioms may be 'formalized' so as to realize Leibniz's dream of finding truth by calculation. The missing ingredient is *formal logic*. The first steps were taken by Boole (1847), in what later became known as Boolean algebra or propositional logic. Boole noticed that the connectives 'or' and 'and' act on propositions rather like sum and product. They satisfy certain basic identities, such as $a + b = b + a$ and $ab = ba$ from which general identities between compound propositions may be proved by algebra.

But propositional logic is not expressive enough for mathematics, where the internal structure of a proposition is important. Typically, a mathematical proposition contains:

Variables ranging over some domain of individuals, such as numbers.
Predicate symbols denoting properties or relations on the domain.
Logic symbols which include not only connectives, such as 'and', 'or', and 'not', but also the quantifiers 'for all x' and 'there exists an x', applied to any variable x.

When these linguistic elements are included we have the language of *predicate logic*. For predicate logic it is not at all clear how to prove the valid propositions – that is, those true for all domains and all interpretations of the predicate symbols – but, amazingly, it is possible. Frege in 1879 gave a set of axioms and rules of inference capable of generating them.

We will not list all of Frege's axioms and rules here. Examples of his axioms, written in terms of the connectives $\Rightarrow$ (if ... then)[1] and $\neg$ (not) and the quantifier $\forall x$ (for all x), are

- $a \Rightarrow (b \Rightarrow a)$,
- $a \Rightarrow (\neg\neg a)$,
- $\forall x P(x) \Rightarrow P(c)$, where c is a letter not in $P(x)$.

Examples of his rules of inference are

- modus ponens: from B and $B \Rightarrow A$ infer A
- $\forall$ introduction: from $A(x)$ infer $\forall x A(x)$.

---

[1] Equivalent to the horseshoe symbol $\supset$ preferred by philosophers.

Frege apparently believed that his axioms and rules of inference suffice to prove any valid proposition. Gödel (1930) proved that Frege was correct. This completeness theorem for predicate logic was the first of several astonishing contributions of Gödel to mathematical logic (and to the philosophy of mathematics).

Gödel's proof of the completeness theorem incidentally proved two other important properties of predicate logic.

Modelling consistent sentences. If $\sum$ is a set of propositions whose logical consequences include no contradiction then there is a model of $\sum$. That is, there is a domain D of individuals, and interpretations of the predicate symbols of $\sum$ on D, under which each proposition in $\sum$ is true.

Compactness. If each finite subset $\sum'$ of $\sum$ has a model, then so has $\sum$. (This follows from modelling, because if each finite subset $\sum'$ has a model, then no contradiction follows from $\sum'$. But then no contradiction follows from $\sum$, since any proof is finite and hence involves only a finite subset $\sum'$ of $\sum$.)

8.4 Completeness and Incompleteness

Predicate logic completeness means that all mathematical theorems are provable in the following relative sense. If τ follows from axioms $\alpha_1, \ldots, \alpha_k$, then the implication $(\alpha_1 \text{ and} \ldots \text{and } \alpha_k) \Rightarrow \tau$ is valid and hence provable in predicate logic. (This calls to mind the opening words of Russell [1903]: pure mathematics is the class if all propositions of the form 'p implies q'.)

However, there may be some area of mathematics that cannot be completely axiomatized, in the sense that no set of axioms we can write down will logically imply all the truths of that area. This actually happens, as we will see in the next section.

8.5 Philosophical Issues

When formal systems were first introduced, many of the questions about them concerned the meaning of the formulas and the methods used to derive theorems, such as induction.

Intuition and logic. Poincaré in 1894 made a forceful contrast between induction in science and in mathematics:

> Induction applied to the physical sciences is always uncertain, because it rests on the belief in a general order of the universe, and order outside us. Mathematical induction … on the contrary, imposes itself necessarily because it is only the affirmation of a property of the mind itself. (see Ewald 1996, II:980)

By describing induction as a property of the mind, he is claiming that induction belongs to intuition, rather than logic. Be that as it may, induction can certainly be expressed as a property of numbers.

Given the reduction of mathematics to arithmetic plus some set theory, it remains to find suitable axioms for arithmetic and sets. For arithmetic, induction seems both intuitive and crucial – but is it sufficient? For sets, it is not obvious which axioms are intuitive or sufficient. How sure are we that the accepted axioms are consistent?

Meaning and existence. Is the existence of infinite sets provable? Bolzano (1851) and Dedekind (1888) thought so, as we saw in Section 7.5, but Zermelo took it as an axiom. On the other hand, a workable theory of a finite mathematical universe does not seem to exist either. If we omit Zermelo's axiom of infinity from his set theory (or add its negation) we obtain a theory of finite sets, but still there are infinitely many finite sets. So, not surprisingly, we cannot find any finite model of the theory of finite sets.

Another question about existence was raised by Hilbert in 1900. I quote from the English translation in Hilbert (1902, 448):

> If contradictory attributes be assigned to a concept, I say, that mathematically the concept does not exist. So, for example, a real number whose square is −1 does not exist mathematically. But if it can be proved that the attributes assigned to the concept can never lead to a contradiction by the application of a finite number of logical processes, I say that the mathematical existence of the concept (for example, of a number or a function which satisfies certain conditions) is thereby proved.

A rather satisfying justification of Hilbert's claim is given by the modelling of consistent sentences that follows from Gödel's proof of the completeness theorem for predicate logic. If the 'attributes assigned to the concept' can be expressed in predicate logic (which is normally the case in mathematics) then the concept has a model, which can serve to establish its existence.

Discrete and continuous. The downside of the model of consistent sentences obtained from Gödel's completeness proof is that it is not necessarily the intended model. In particular, if the set of sentences is countable – which for all practical purposes it must be – then the model is also countable. In particular any consistent set of sentences about the continuum has a countable model, which is certainly not intended! This result, due to Skolem (1922) and known as the Skolem paradox, is not actually contradictory. It means only that there is no function in the model that pairs the members of its 'continuum' with natural numbers. But it certainly shows how hard it is to completely describe the continuum.

Some mathematicians objected to formalization, not only because they did not believe in some of the objects being formalized, but also on the grounds that formalization excludes intuition, which they considered to be a vital ingredient in mathematics. The latter objection could have been quashed, in principle, if a complete formal system for mathematics were found – but it won't be, as we will see in the next section. In any case the objection was beside the point.

Even the strongest advocates of formalization, such as Hilbert, valued the role of intuition in mathematics. Their reason for formalization was otherwise: to produce proofs that everyone agrees *are* proofs. Everyone will agree that a given theorem follows from given axioms by given rules, because this is a statement about a finite computation that anyone can verify. The interpretation of the strings of symbols in the proof is beside the point. The point of formalization is, rather, to make the question of consistency a question about computation; namely, whether a contradictory string such as $0 = 1$ follows from the axioms by the given rules. Once this question is answered affirmatively for a formal system $\mathcal{F}$ then even constructivists will have to admit that $\mathcal{F}$ is harmless, even if they do not consider all its theorems to be meaningful.

This was the Hilbert programme, with which he hoped to prove that reasoning about infinity is harmless. The programme did not turn out as Hilbert hoped, as we will see in the next section, but it did lead to remarkable developments in the philosophy of mathematics.

9 Unsolvability and Incompleteness

PREVIEW

The formal systems discussed in the previous section are a partial realization of Leibniz's dream of a calculus ratiocinator – a formal system for establishing truth – but only partial. In this section we look at the other side of the story: the limitations of formal systems. The limitations stem from a stunning proposition known as *Church's thesis*: there is a complete and mathematically precise definition of computation (and, with it, precise definitions of algorithm, solvable problem, and unsolvable problem).

The possibility of an all-embracing definition of computation was first suspected by Post in the 1920s, and most convincingly argued by Turing (1936). By its nature, Church's thesis can in principle be falsified, but no falsification has ever come to light. So Church's thesis is now considered to be as well confirmed as any law of nature.

At any rate, we can build a theory of computation on Church's thesis, and this theory leads to the following remarkable results:

1 There are algorithmically unsolvable problems in mathematics.
2 Any sufficiently strong formal system for arithmetic is incomplete or inconsistent.
3 The consistency of a sufficiently strong formal system for arithmetic is not provable within the system.

These results, as we will show in outline below, all follow from simple variations on Cantor's diagonal argument.

9.1 Computability

In the seventeenth century, when Leibniz dreamed of deciding truth by computation, the concept of computation had a rather limited meaning. Leibniz himself had designed a computing machine capable of doing arithmetic on numbers, and no doubt he would have accepted that algebra was computation too. By 1850 Boole had got as far as doing propositional logic by algebraic computation. But a general definition of computation had to wait for the development of formal systems for mathematics, around 1900. Only then did it become clear how broad the definition of computation needed to be in order to make Leibniz's dream come true.

The most influential formal system in the early twentieth century was the *Principia Mathematica* of Whitehead and Russell (1910). The *Principia* claimed to show how all theorems of mathematics could be generated from particular axioms by certain rules. The rules were such that, in principle, they were mechanical and hence could be applied without thought to strings of symbols – eliminating all possibility of human error or bias. In the early 1920s the rules were analysed and simplified by Post, until they were reduced to the form

$$gW \rightarrow Wg',$$

for a finite number of pairs of symbol strings (g, g'). The rule $gW \rightarrow Wg'$ says that, in any string beginning with g, the g may be removed from the left and g' then attached on the right. Post called such a system of rules a *normal system*.

Post thought at first that the simplicity of normal systems would enable him to decide whether a given string of symbols was a theorem of *Principia* or not. But then he found himself unable to predict the behaviour of very simple normal systems, and came to the realization that the situation was the opposite of what he had hoped: any computation can be simulated by a normal system, and there is no general rule for deciding whether a given normal system produces a given string.

Post had glimpsed the future of mathematical logic as it was to unfold over the next fifteen years (for his account, see Post 1941). However, he was held back by an unprecedented difficulty: how can one be sure that the concept of normal system (or any other definition) completely captures the concept of computation? He saw that this claim (later known as Church's thesis or the Church–Turing thesis) is something like a law of nature – one in need of continual verification, and at risk of possible falsification.

9.2 Unsolvability

While Post's work remained unpublished and unknown, independent attempts to define the notion of computation were made by Church (1936) and Turing (1936). Turing's approach (now known as the *Turing machine* concept) was remarkably convincing, being basically an idealization of a human 'computer' working with pencil and paper:

- Instead of paper, a Turing machine has an infinite tape divided into squares. Each square can hold one from a finite alphabet of symbols, including the blank.
- Instead of the human with a pencil, a Turing machine has a read/write head that can assume one of a finite number of internal states (like mental states). The head scans one square at a time and, depending on the internal state q_i and scanned symbol S_j, replaces S_j by a symbol S_k, moves one square to the left or right, and enters a state q_l.

Each Turing machine is therefore described by finitely many quintuples, of the form either $q_i S_j S_k R q_l$ or $q_i S_j S_k L q_l$. A computation of the machine is determined by an input, consisting of a finite sequence of marked tape squares, the square initially scanned, and the initial state.

With a little practice it is easy to realize pencil-and-paper computations by Turing machine computations. The secret is to imagine how the computation could be done if one is allowed to view only one symbol at a time. The most ambitious computation one needs to think about is that of a universal Turing machine: a machine U that can take the description of an arbitrary Turing machine T, and arbitrary input I, and simulate the computation of T on I.

The first difficulty to overcome is that U, like any other Turing machine, has only a finite alphabet of symbols and finitely many internal states. If U is even to read the description of T that description must be written in a fixed finite alphabet. For example, one could use the alphabet $\{q,S,L,R,{}'\}$, and use the prime symbol $'$ to rewrite q_i as $q''^{\cdots}$ (q with i primes) and to rewrite S_j as $S''^{\cdots}$ (S with j primes). Thus a single symbol in the description of T must be replaced by a string of symbols, necessarily spread over a sequence of squares of U's tape.

Naturally, this makes U's simulation of T rather slow (as does the need for U to continually 'refer back' to the description of T in order to carry out each step of T's computation). Nevertheless, one sees in principle why a universal Turing machine exists, and that U can simulate what T does on a given input, step by step. (Today, universal Turing machines are ubiquitous; any common programming language is equivalent to a universal Turing machine. The downside of this fact, as we are about to prove, is that it is hard to foresee what a given programme will do on a given input.)

Now it is one thing to follow instructions; it is another to foresee where they will lead. For example, we can compute any number of decimal places of π, but at present we do not know whether 1000 consecutive 7s will ever occur. For Turing machines the ultimate outcome of computations is provably uncertain, as we can see with the following:

Self-examination problem. Given a Turing machine T decide whether T, with its own description des (T) as input, ever halts on a blank square.

We are using the term 'problem' here to mean an infinite set of questions, in this case the following, for each Turing machine T:

Q_T: Does T, on input des (T), ever halt on a blank square?

We are going to show that no Turing machine S can correctly answer all the questions Q_T. It is fair to assume that S receives question Q_T in the form des (T), because Q_T can be reconstructed from des (T). It is also fair to assume that S answers 'no' by halting on a blank square, and 'yes' by halting on a non-blank square.

But then S cannot give the correct answer to question Q_S. If S, given input Q_S, halts on a blank square, then the answer to Q_S is 'yes', so S should *not* halt on a blank square. And if S does not halt on a blank square then the answer to question Q_S is 'no', and S must halt on a blank square.

This contradiction shows that the self-examination problem cannot be solved by Turing machine. And therefore, if the Church–Turing thesis is correct, this problem cannot be solved by any computation whatever. As we say, the problem is *algorithmically unsolvable* or, simply, unsolvable.

Since the self-examination problem is rather obviously self-defeating, one might hope that unsolvability is an aberration, not something that happens naturally. This is not so, because Turing machines can be 'simulated' by various natural systems in mathematics and logic. In fact Church and Turing both noticed immediately that predicate logic can 'simulate' Turing machines, with the result the problem of deciding validity in predicate logic – the so-called *Entscheidungsproblem* – is unsolvable.

9.3 Incompleteness

The notion of computability, which by some miracle seems to be complete and absolute, stands in contrast to the notion of provability, which turns out to be incomplete and relative. The link between the two is *non*-computability, which follows from computability by the diagonal argument. The most convenient form of the diagonal argument is the one used by Cantor (1891) to prove that any set has more subsets than elements. Following Post, we apply this argument to sets associated with Turing machines, called computably enumerable sets.

A set of natural numbers is called *computably enumerable* if there is a Turing machine that lists its elements. The manner of making the list is not important, as long as any Turing machine has a computably enumerable set associated with it (possibly the empty set), and we can observe when a given machine T lists a given number n. We will appeal to the Church–Turing thesis to claim that any set that is intuitively listable is listable by Turing machine. $\mathbb{N}$ is computably enumerable, and so are many of its subsets, such as the set of prime numbers. Moreover, we can computably enumerate all the Turing machine descriptions, by listing them in lexicographical order. We let W_n be the computably enumerable set listed by the n th Turing machine.

It follows by the diagonal argument that the set

$$D = \{n : n \notin W_n\}$$

is not computably enumerable, because it differs from each W_n with respect to the element n. (Interestingly, its complement $\mathbb{N} - D = \{n : n \in W_n\}$ is computably enumerable. We can list the n that appear in $\mathbb{N} - D$ by allowing the first k Turing machines to run for k steps, for $k = 1, 2, 3, \ldots$, and putting n on the list if ever it is listed by the n th machine.)

Now a formal system $\mathcal{F}$, by the broadest possible definition, has a computably enumerable set of theorems, so we can computably enumerate the theorems of $\mathcal{F}$ that have the form $n \notin D$. Since D is not computably enumerable, these theorems (if correct) do not include all true statements of the form $n \notin D$. This fact points to some failure of the system $\mathcal{F}$, but we have to tease out the significance of the proviso 'if correct'. This leads to a sequence of increasingly strong incompleteness theorems.

1 Suppose that $\mathcal{F}$ is consistent and that the set of n for which $\mathcal{F}$ proves '$n \notin W_n$' is the computably enumerable set W_m. We also assume that $\mathcal{F}$ is strong enough to simulate Turing machines, and hence to prove $n \in W_n$ whenever this is true. Now what can we say about the sentence '$m \notin W_m$'?

If $\mathcal{F}$ proves '$m \notin W_m$', then $m \in W_m$, by definition of W_m. But we have assumed that $\mathcal{F}$ can prove all such true statements, so $\mathcal{F}$ also proves '$m \in W_m$', contradicting consistency. Thus $\mathcal{F}$ does *not* prove '$m \notin W_m$', so $m \notin W_m$, by the definition of W_m again. This makes '$m \notin W_m$' a specific true sentence that $\mathcal{F}$ fails to prove.

Notice that the sentence '$m \notin W_m$' essentially says 'I am not provable', because $m \notin W_m$ means '$m \notin W_m$' is not provable, by definition of W_m.

2 By translating the workings of Turing machines into arithmetic – which is possible though far from obvious – we can show that Peano Arithmetic PA (see Section 8.2) is 'strong enough' in the above sense. Thus the language of PA has a sentence that is equivalent to $m \notin W_m$, and hence unprovable in PA, if PA is consistent. In other words, if PA is consistent, then the sentence '$m \notin W_m$' is not provable in PA. Or, equivalently, if PA is consistent, then $m \notin W_m$.

3 Similar 'arithmetization' of proofs in PA lets us express consistency of PA by a sentence in the language of PA, Con(PA), and to deduce from it that $m \notin W_m$. So 'Con(PA) $\Rightarrow m \notin W_m$' is a theorem of PA.

It follows that Con(PA) cannot be proved in PA, otherwise modus ponens would give a proof of '$m \notin W_m$', and we know that '$m \notin W_m$' has no proof in PA. Thus PA, if consistent, *cannot prove its own consistency* (and neither can any consistent system that includes PA, because a similar argument will apply).

This is a brilliant train of thought – one of the most stunning in mathematics. Item 1 resembles the results found by Post in the 1920s. Items 2 and 3 are essentially the first and second incompleteness theorems of Gödel (1931), though Gödel found the first theorem differently, by directly constructing a sentence that says 'I am not provable'. Interestingly, Gödel first proved incompleteness for the higher-level system of *Principia Mathematica* because he did not realize that computation or proof could be arithmetized. He was prompted to arithmetize by von Neumann, who also deserves some credit for the second incompleteness theorem. Von Neumann pointed out unprovability of consistency in a letter to Gödel before Gödel announced the result himself.

9.4 The Incompleteness of Set Theory

The diagonal argument is an easy and powerful way to demonstrate the incompleteness of common axiom systems for mathematics such as Peano arithmetic. But it has so far failed to show the unprovability (in PA or any stronger system) of any well known unproved sentences, such as the Goldbach conjecture, twin prime conjecture, or Riemann hypothesis. In fact, all of the unprovable sentences found so far are ones devised by logicians.

The situation is different in set theory, where several propositions open since Cantor's time were subsequently shown to be unprovable from the standard set theory axioms. Two of the most interesting are the axiom of choice and the continuum hypothesis.

The axiom of choice states that any set X of non-empty sets x has a choice function: a function f such that $f(x) \in x$ for each $x \in X$. This axiom is used implicitly (and in the early days it was often used unconsciously) whenever a set is defined by means of an infinite sequence of choices. For example, to prove that an infinite set X contains an infinite sequence of elements $x_1, x_2, x_3, \ldots$ one wants to say: choose x_1 from X, then x_2 from $X - \{x_1\}$, then x_3 from $X - \{x_1, x_2\}$, and so on. But even this simple definition cannot be justified by the standard axioms of set theory. Cohen (1963a,b) showed this by constructing a model of the standard axioms in which there is an infinite set (of real numbers) containing no infinite sequence.

The continuum hypothesis arose from Cantor's discovery that $\mathbb{R}$ is a set larger than $\mathbb{N}$. In its naive form, the hypothesis states that $\mathbb{R}$ is the next largest set after $\mathbb{N}$. (Cantor also gave a more sophisticated form, involving ordinal numbers, that I do not have space to explain here.) Cohen (1963a,b) showed that the continuum hypothesis is not provable from the standard axioms, by constructing a model of these axioms in which the hypothesis is false. It should also be mentioned that the axiom of choice and the continuum hypothesis cannot be *disproved* from the standard axioms. This was shown by Gödel (1938) when he constructed a model in which both statements hold.

Thus two of the most natural questions about infinite sets cannot be settled by the natural axioms. The situation is a little like the situation of the parallel axiom relative to the other axioms of Euclid (or Hilbert). We are free to add either axiom or its negation and no contradiction will arise, assuming that the other axioms are consistent. The difference is that, as far as we know, for set theory there are no obviously natural models for the alternatives – no 'Cantorian' and 'non-Cantorian' set theories.[2]

---

[2] Perhaps it would be better to say that the only known models for 'Cantorian' and 'non-Cantorian' set theories are those constructed with the purpose of modelling these theories. This contrasts with

9.5 Philosophical Issues

Taking the developments from the 1930s to the 1960s into account, we can see how far the philosophy of mathematics has evolved since ancient Greece:

1 The concept of computability, unrecognized by the Greeks, has been precisely defined and now rules large areas of mathematics. It sets the limits for the concept of proof, and adjudicates whether many problems are solvable or not.

2 The concept of proof in logic has been described completely (in the case of predicate logic, the one most relevant to mathematics): there is a computation that generates all logically valid formulas. However, there is no algorithm that decides, given an arbitrary formula, whether that formula is provable (hence valid) or not.

3 The concept of proof in mathematics is incomplete, and hence falls short of absolute mathematical truth; in fact, for any consistent system $\mathcal{F}$ containing a certain amount of arithmetic there are true statements of $\mathcal{F}$ not provable by $\mathcal{F}$. On the bright side, it follows from the completeness of logic that we can find all consequences of a given set of axioms.

4 A claim of Hilbert, that mathematical existence is the same as consistency, has been found tenable, since Gödel's proof of the completeness of predicate logic gives a model for any consistent set of axioms (as mentioned in Section 8.3). However, a computable set of axioms need not have a computable model, so this kind of existence is not acceptable to constructivists.

5 The consistency of mathematics has to be a matter of intuition; in the sense that any reasonably strong (and consistent) system contains no proof of its own consistency. The Hilbert program for proving consistency seems damaged beyond repair by this result.

6 The meaning and existence of infinity has become a more complicated question. The continuum, which most mathematicians accept, can be grasped only as an actual infinity, because it is uncountable. But there are infinitely many actual infinities. It is unclear where to draw the line between acceptable and unacceptable in the actual infinities.

7 Despite its difficulties, the continuum has become the basis for analysis, hence most of geometry, and most of mathematical physics. These fields of mathematics have been arithmetized; that is, based on axioms for the natural numbers plus certain axioms for infinite sets.

---

the situation in geometry, where non-Euclidean geometry was found to hold in certain structures that had been studied before there was a suitable geometric language to describe them. We mentioned this in section 6.3.

8 The continuum is still not completely understood, since the continuum hypothesis is not settled by the standard axioms of set theory. On the other hand, no 'constructive' analogue of the continuum has been found that is acceptable to the majority of mathematicians.

Bibliography

Beltrami, E. (1868). Teoria fondamentale degli spazii di curvatura costante. *Annali di Matematica Pura ed Applicata, Ser. 2* 2, 232–55. In his *Opere Matematiche 1*, 406–29, English translation in Stillwell (1996).

Bolzano, B. (1817). *Rein analytischer Beweis des Lehrsatzes dass zwischen je zwey Werthen, die ein entgegengesetzes Resultat gewähren, wenigstens eine reelle Wurzel der Gleichung liege.* Ostwald's Klassiker 153. Repr., Leipzig: Engelmann, 1905. English translation in Russ (2004), 251–77.

Bolzano, B. (1851). *Paradoxien der Unendlichen.* Repr., Leipzig: Bei C. H. Reclam Sen. English translation in Russ (2004), 591–678.

Bombelli, R. (1572). *L'algebra. Prima edizione integrale. Introduzione di U. Forti. Prefazione di E. Bortolotti.* Repr., Milan: Giangiacomo Feltrinelli Editore LXIII, 1966.

Boole, G. (1847). *Mathematical Analysis of Logic.* Repr., London: Basil Blackwell, 1948.

Cantor, G. (1874). Über eine Eigenschaft des Inbegriffes aller reellen algebraischen Zahlen. *Journal für reine und angewandte Mathematik 77*, 258–62. In his *Gesammelte Abhandlungen*, 145–8. English translation by W. Ewald in Ewald (1996), 2:840–43.

Cantor, G. (1891). Über eine elementare Frage der Mannigfaltigkeitslehre. *Jahresbericht deutschen Mathematiker-Vereinigung 1*, 75–8. English translation by W. Ewald in Ewald (1996), 2:920–22.

Cardano, G. (1545). *Ars magna.* Translation in Richard Witmer, *The Great Art or the Rules of Algebra*, Cambridge, MA: MIT Press, 1968.

Church, A. (1936). An unsolvable problem in elementary number theory. *American Journal of Mathematics 58*, 345–63.

Cohen, P. (1963a). The independence of the continuum hypothesis I. *Proceedings of the National Academy of Sciences of the United States of America 50*, 1143–8.

Cohen, P. (1963b). The independence of the continuum hypothesis II. *Proceedings of the National Academy of Sciences of the United States of America 51*, 105–10.

Davis, M. (Ed.) (2004). *The Undecidable.* Mineola, NY: Dover Publications Inc. Corrected reprint of the 1965 original [Raven Press, Hawlett, NY].

Dedekind, R. (1888). *Was sind und was sollen die Zahlen?* Braunschweig: Vieweg und Sohn. English translation in *Essays on the Theory of Numbers*, New York: Dover, 1963.

Dehn, M. (1900). Über raumgleiche Polyeder. *Göttingen Nachrichten 1900*, 345–54.

Descartes, R. (1637). *The Geometry of René Descartes (With a Facsimile of the First Edition, 1637)*. Translated by David Eugene Smith and Marcia L. Latham. Repr., New York: Dover, 1954.

Euler, L. (1748). *Introductio in analysin infinitorum, I*. Volume 8 of his *Opera Omnia*, series 1. English translation by John D. Blanton, *Introduction to the Analysis of the Infinite*, Book I, New York: Springer, 1988.

Ewald, W. (1996). *From Kant to Hilbert: A Source Book in the Foundations of Mathematics*. Vols. I and II. Oxford, UK: Clarendon Press.

Field, H. H. (1980). *Science without Numbers*. Princeton, NJ: Princeton University Press.

Frege, G. (1879). *Begriffschrift*. English translation in van Heijenoort (1967), 5–82.

Gauss, C. F. (1816). Demonstratio nova altera theorematis omnem functionem algebraicum rationalem integram unius variabilis in factores reales primi vel secundi gradus resolvi posse. *Commentationes societas regiae scientiarum Gottingensis recentiores* 3, 107–42. In his Werke 3, 31–56.

Gödel, K. (1930). Die Vollständigkeit der Axiome des logischen Funktionenkalküls. *Monatshefte für Mathematik und Physik 37*, 349–60. English translation in Gödel (1986), 103–23.

Gödel, K. (1931). Über formal unentscheidbare Sätze der *Principia Mathematica* und verwandter Systeme. I. *Monatshefte für Mathematik und Physik 38*, 173–98. English translation in van Heijenoort (1967), 596–616.

Gödel, K. (1938). The consistency of the axiom of choice and the generalized continuum hypothesis. *Proceedings of the National Academy of Sciences 25*, 220–24.

Gödel, K. (1986). *Collected Works. Vol. I*. The Clarendon Press, Oxford University Press, New York. Publications 1929–1936, Edited and with a preface by Solomon Feferman.

Grassmann, H. (1844). *Die lineale Ausdehnungslehre*. Otto Wiegand, Leipzig. English translation in Grassmann (1995), 1–312.

Grassmann, H. (1847). *Geometrische Analyse geknüpft an die von Leibniz gefundene Geometrische Charakteristik*. Weidmann'sche Buchhandlung, Leipzig. English translation in Grassmann (1995), 313–414.

Grassmann, H. (1861). *Lehrbuch der Arithmetic*. Berlin: Enslin.

Grassmann, H. (1995). *A New Branch of Mathematics*. Chicago: Open Court. The Ausdehnungslehre of 1844 and other works, translated from the German with a note by Lloyd C. Kannenberg, and with a foreword by Albert C. Lewis.

Hamilton, W. R. (1835). Theory of conjugate functions, or algebraic couples. Communicated to the Royal Irish Academy, 1 June 1835. *Mathematical Papers 3*: 76–96.

Heath, T. L. (1956). *The Thirteen Books of Euclid's Elements translated from the text of Heiberg. Vol. I: Introduction and Books I, II. Vol. II: Books III–IX. Vol. III: Books X–XIII and Appendix.* Translated with introduction and commentary by Thomas L. Heath. 2nd edn. New York: Dover.

Hilbert, D. (1899). *Grundlagen der Geometrie.* Leipzig: Teubner. English translation in Foundations of Geometry, Chicago: Open Court, 1971.

Hilbert, D. (1902). Mathematical problems. *Bulletin of the American Mathematical Society 8*, 437–79. Translated by Frances Winston Newson.

Huygens, C. (1659). Fourth part of a treatise on quadrature. *Œuvres Complètes* 14: 337.

Lambert, J. H. (1766). Die Theorie der Parallellinien. *Magazin für reine und angewandte Mathematik 1786*, 137–64, 325–58.

Minkowski, H. (1908). Raum und Zeit. *Jahresbericht der Deutschen Mathematiker-Vereinigung 17*, 75–88.

Peano, G. (1888). *Calcolo Geometrico secondo l'Ausdehnungslehre di H. Grassmann, preceduto dalle operazioni della logica deduttiva.* Turin: Bocca. English translation in Peano (2000).

Peano, G. (1889). *Arithmetices principia.* Turin: Bocca.

Peano, G. (2000). *Geometric Calculus.* Boston, MA: Birkhäuser. According to the Ausdehnungslehre of H. Grassmann, translated from the Italian by Lloyd C. Kannenberg.

Poincaré, H. (1952). *Science and Method.* New York: Dover. Translated by Francis Maitland, with a preface by Bertrand Russell.

Post, E. L. (1941). Absolutely unsolvable problems and relatively undecidable propositions – an account of an anticipation. In Davis (2004), 338–433.

Riemann, G. F. B. (1854). Über die Hypothesen, welche der Geometrie zu Grunde liegen. In *Werke*, 2nd edn., 272–87. English translation in Ewald (1996), 2:652–61.

Russ, S. (2004). *The Mathematical Works of Bernard Bolzano.* Oxford: Oxford University Press.

Russell, B. (1903). *The Principles of Mathematics.* Cambridge: Cambridge University Press.

Saccheri, G. (1733). *Euclid Vindicated from Every Blemish.* Classic Texts in the Sciences. Repr., Cham, Switzerland: Birkhäuser/Springer, 2014. Dual Latin–English text, edited and annotated by Vincenzo De Risi translated from the Italian by G. B. Halsted and L. Allegri.

Schwarz, H. A. (1872). Über diejenigen Fälle, in welchen die Gaussische hypergeometrische Reihe eine algebraische Function ihres vierten Elementes darstellt. *Journal für reine und angewandte Mathematik* 75, 292–335. In his Mathematische Abhandlungen 2, 211–259.

Skolem, T. (1922). Einige Bemerkungen zu axiomatischen Begründung der Mengenlehre. In *Mathematikerkongressen i Helsingfors den 4–7 Juli 1922, Den femte skandinaviska matematikerkongressen, Redogörelse*, 217–32. English translation in van Heijenoort (1967), 290–301.

Stillwell, J. (1996). *Sources of Hyperbolic Geometry*. Providence, RI: American Mathematical Society.

Turing, A. (1936). On computable numbers, with an application to the *Entscheidungsproblem*. *Proceedings of the London Mathematical Society 42*, 230–65.

van Heijenoort, J. (1967). *From Frege to Gödel: A Source Book in Mathematical Logic, 1879–1931*. Cambridge, MA: Harvard University Press.

Vilenkin, N. Y. (1995). *In Search of Infinity*. Boston, MA: Birkhäuser. Translated from the Russian original by Abe Shenitzer with the editorial assistance of Hardy Grant and Stefan Mykytiuk.

von Koch, H. (1904). Sur une courbe continue sans tangente, obtenue par une construction géométrique élémentaire. *Archiv för Matematik, Astronomi och Fysik 1*, 681–704.

Whitehead, A. N., and B. Russell (1910–13). *Principia Mathematica.* 3 vols. Cambridge: Cambridge University Press.

Zermelo, E. (1908). Untersuchungen über die Grundlagen der Mengenlehre I. *Mathematische Annalen 65*, 261–81. English translation in van Heijenoort (1967), 200–215.

Cambridge Elements ≡

The Philosophy of Mathematics

Penelope Rush

University of Tasmania

From the time Penny Rush completed her thesis in the philosophy of mathematics (2005), she has worked continuously on themes around the realism/anti-realism divide and the nature of mathematics. Her edited collection *The Metaphysics of Logic* (Cambridge University Press, 2014), and forthcoming essay 'Metaphysical Optimism' (*Philosophy Supplement*), highlight a particular interest in the idea of reality itself and curiosity and respect as important philosophical methodologies.

Stewart Shapiro

The Ohio State University

Stewart Shapiro is the O'Donnell Professor of Philosophy at The Ohio State University, a Distinguished Visiting Professor at the University of Connecticut, and a Professorial Fellow at the University of Oslo. His major works include *Foundations without Foundationalism* (1991), *Philosophy of Mathematics: Structure and Ontology* (1997), *Vagueness in Context* (2006), and *Varieties of Logic* (2014). He has taught courses in logic, philosophy of mathematics, metaphysics, epistemology, philosophy of religion, Jewish philosophy, social and political philosophy, and medical ethics.

About the Series

This Cambridge Elements series provides an extensive overview of the philosophy of mathematics in its many and varied forms. Distinguished authors will provide an up-to-date summary of the results of current research in their fields and give their own take on what they believe are the most significant debates influencing research, drawing original conclusions.

Cambridge Elements$\equiv$

The Philosophy of Mathematics

Elements in the Series

Mathematical Structuralism
Geoffrey Hellman and Stewart Shapiro

A Concise History of Mathematics for Philosophers
John Stillwell

A full series listing is available at: www.cambridge.org/EPM

Made in the
USA
Middletown, DE